Learn College Algebra NOW!

Minute Help Guides

Minute Help Press

www.minutehelp.com

Table of Contents

Introduction

Algebra is a pretty important tool for every other area of mathematics, which is why it comes first, after you learn arithmetic. It's the basic setting you use to handle the unknown, putting a situation into mathematical terms and using various strategies to figure out what's going on. And if you're having problems with it, that's okay—we all have our own specialties, and there's not an area in any subject that everyone is great at. That's what this brief course is for: to give you a hand where you're facing difficulties with algebra. With that in mind, let's begin.

Chapter 1: Equations and Inequalities

Equations

Basically, an equation is a mathematical statement that asserts the equality of two expressions. In simpler terms, an equation says "Hey, look at these two things. They have the same value." For now, we will be dealing with the simple case of equations like $x + 3 = 5$, where there is only one value for x that will make the statement true.

The most important thing you must remember when solving an equation or inequality is that you must always do the same thing to both sides—add or subtract the same term, multiply or divide by the same term. After all, if two things are the same, and one of them changes, then they aren't the same any more.

> *Note: Normally, x is used to show multiplication—however, since x is being used as a variable here, that doesn't work so well, so it's being replaced with . If you see •, that means "times".*
> *4 x 6 is the same as 4 • 6 and the same as 24.*

The basic idea when solving an equation is to get x by itself so that it is the only thing on one side of the equation. If x is being multiplied by some number, having another added to it, and that combination is equal to a third, then the whole thing can be written as $ax + b = c$. c, b, and a may be numbers like 2 or 3, or they might be $\frac{-2}{3}$ or $3\frac{5}{7}$. It doesn't really make any difference when you get right down to it, because no matter what, if $ax + b = c$, then you simply subtract b from both sides and divide the results by a: $x = \frac{c - b}{a}$. By the way, in this notation, a is called the coefficient and b is the slope intercept. c doesn't get a name, and we'll see why when we cover coordinates and graphs in the next section.

Example: Solve $2x - 3 = 7$.

Solution:

Step 1: Isolate x by adding 3 to both sides.

$2x - 3 = 7$

$(2x - 3) + 3 = (7) + 3$

$2x = 10$

Step 2: Finish isolating x by dividing both sides by 2.

$\frac{2x}{2} = \frac{10}{2}$

$x = 5$

4

Answer: $x = 5$

Example: Solve $\frac{x}{3} + 4 = 7$.

Solution:

Step 1: Isolate x by subtracting 4 from both sides.

$(\frac{x}{3} + 4) - 4 = (7) - 4$

$\frac{x}{3} = 3$

Step 2: Finish isolating x by multiplying both sides by 3.

$\frac{x}{3} * 3 = 3 * 3$

$x = 9$

Answer: $x = 9$

Example: Solve $4x - 3 = 2x + 1$.

Solution:

Step 1: Restrict x to one side of the equation by subtracting 2x from both sides.

$(4x - 3) - 2x = (2x + 1) - 2x$

$2x - 3 = 1$

Step 2: Isolate x by adding 3 to both sides.

$(2x - 3) + 3 = (1) + 3$

$2x = 4$

Step 3: Finish isolating x by dividing both sides by 2.

$\frac{2x}{2} = \frac{4}{2}$

$x = 2$

Answer: $x = 2$

Inequalities

The concept of an inequality is closely related to an ordinary equation and adds several types of relationships between expression A and expression B. A might be less than B, for example, or A might be greater than or equal to B. Just to refresh your memory, < is less than and > is greater than—the big end of the arrow is always near the bigger number or expression, like an alligator's wide-open mouth approaching the biggest dinner. Also, $\leq$ is less than or equal to, combining $<$ and $=$, and $\geq$ is greater than or equal to, combining $>$ and $=$.

$a < b$	a is less than b: $4 < 6$
$a > b$	a is greater than b: $7 > 3$
$a \leq b$	a is less than or equal to b: $4 \leq 4$
$a \geq b$	a is greater than or equal to b: $5 \geq 4$

The rest we'll explain by showing it to you.

Example: Solve $\dfrac{x}{5} - 7 < 3$.

Solution:

Step 1: Get x alone by adding 7 to both sides.

$(\dfrac{x}{5} - 7) + 7 < (3) + 7$

$\dfrac{x}{5} < 10$

Step 2: Finish by multiplying both sides by 5.

$\dfrac{x}{5} \cdot 5 < 10 \cdot 5$

$x < 50$

Answer: $x < 50$

Example: Solve $2x + 5 > 9$.

Solution:

Step 1: Get x alone by subtracting 5 from both sides.

$(2x + 5) - 5 > (9) - 5$

$2x > 4$

Step 2: Finish by dividing both sides by 2.

$\dfrac{2x}{2} > \dfrac{4}{2}$

$x > 2$

Answer: $x > 2$.

Example: Solve $3x - 10 \geq 2$.

Solution:

Step 1: Get x alone by adding 10 to both sides.

$(3x - 10) + 10 \geq (2) + 10$

$3x \geq 12$

Step 2: Finish by dividing both sides by 3.

$\dfrac{3x}{3} \geq \dfrac{12}{3}$

$x \geq 4$

Answer: $x \geq 4$.

Example: Solve $\dfrac{3x}{7} + 2 \leq 4$.

Solution:

Step 1: Begin isolating x by subtracting 2 from both sides.

$(\dfrac{3x}{7} + 2) - 2 \leq (4) - 2$

$\dfrac{3x}{7} \leq 2$

Step 2: Continue isolating by dividing both sides by the coefficient of x, $\dfrac{3}{7}$, which is the same as multiplying by its inverse, $\dfrac{7}{3}$.

$(\dfrac{3x}{7}) \cdot \dfrac{7}{3} \leq (2) \cdot \dfrac{7}{3}$

$x \leq \dfrac{14}{3}$

Answer: $x \leq \dfrac{14}{3}$, or equivalently, $x \leq 4\dfrac{2}{3}$.

There's one more vital principle to remember: multiplying or dividing by negative numbers switch things around.

Example: Solve $-2x + 5 > 17$.

Solution:

Step 1: Get x alone by subtracting 5 from both sides.

$(-2x + 5) - 5 > (17) - 5$

$-2x > 12$

Step 2: Finish by dividing both sides by -2, making sure to flip the ≥ (greater-than sign) to a ≤ (less-than sign).

$-2x > 12$

$\dfrac{-2x}{-2} < \dfrac{12}{-2}$

$x < -6$

Answer: $x < -6$.

Example: Solve $-\dfrac{x}{3} + 2 \leq 14$.

Solution:

Step 1: Begin by subtracting 2 from both sides.

$(-\dfrac{x}{3} + 2) - 2 \leq (14) - 2$

$-\dfrac{x}{3} \leq 12$

Step 2: End by multiplying both sides by -3, making sure to flip the less-than or equal to sign to a greater-than or equal to sign.

$-\dfrac{x}{3} \leq 12$

$(\dfrac{-x}{3}) \cdot -3 \geq (12) \cdot -3$

$x \geq -36$

Answer: $x \geq -36$.

Looking back, there are two important things to remember when solving inequalities instead of equations.

- Your answer changes from a simple statement of, say "$x = 5$", meaning that 5 is the only value that works in the equation, to a range of values. If the solution is "$x > 3$", then any value of x that is greater than 3 works in the equation, though it must be greater than 3 and not exactly 3.

- When multiplying or dividing both sides by a negative number, flip the inequality sign so that it points in the opposite direction.

$-2x < 5$ becomes $x > \dfrac{-5}{2}$

After all, $6 > 5$, but $-6 < -5$.

Recap

When solving an equation of the form $ax + b = cx + d$, the goal is to get x alone, by *combining like terms*. (You'll encounter this phrase later on, especially in other math courses, so remember it.) This will be easier if it's in a simpler form, but the principle remains that you should:

1. Subtract cx from both sides and subtract b from both sides, to get $(a - c)x = d - b$. Then

2. Divide both sides by (a – c) to get $x = \dfrac{d - b}{a - c}$.

Again, this will be easier if c is zero, so you start with $ax + b = d$, but that won't always the case.

The same method applies with solving inequalities, but remember the two things we just covered:

1. You'll be dealing with one of these: $<><\leq\geq$, thus providing a range of solutions, not a single answer.

2. If $a + c$ is less than zero, you'll be multiplying or dividing by a negative number, and you have to flip the sign from $>$ to $<$, or vice versa. If you have $\leq$ or $\geq$, the same thing happens, but you keep the equals - as we said in the beginning of the chapter, if two things are the same and you do the same thing to them, they stay the same.

Chapter 2: Coordinates and Graphs

Equations can also be interpreted visually. To do this, you use a system called Cartesian coordinates. These coordinates define every point in the area you're looking at with a pair of numbers. One number describes the distance to the **origin** left or right along the horizontal x **axis**, and the other is the distance to the origin up or down along the vertical y **axis**. They are written in the order (x, y), which is also called an **ordered pair**. We've covered the first of those here in the previous section, and the other is not much different.

Coordinate Grid

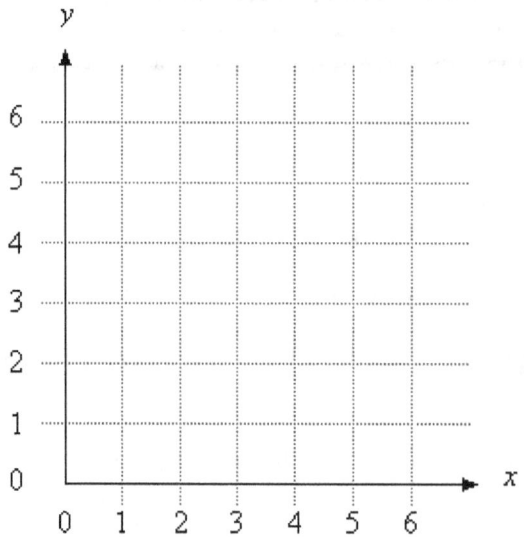

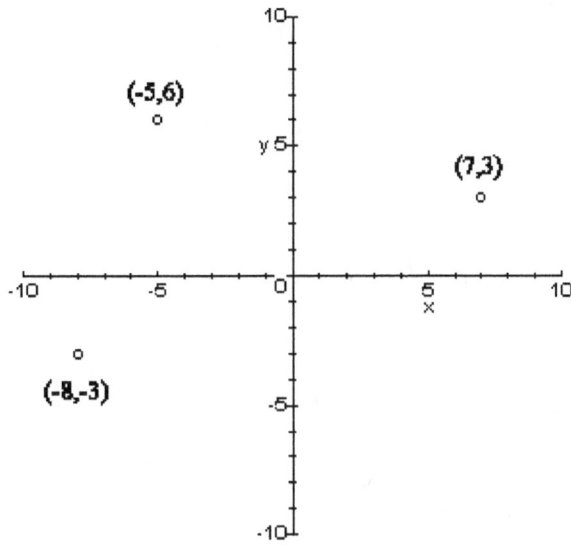

Let's just dive right in and show an example.

Here's a graph of the equation $3x - 4 = 7$.

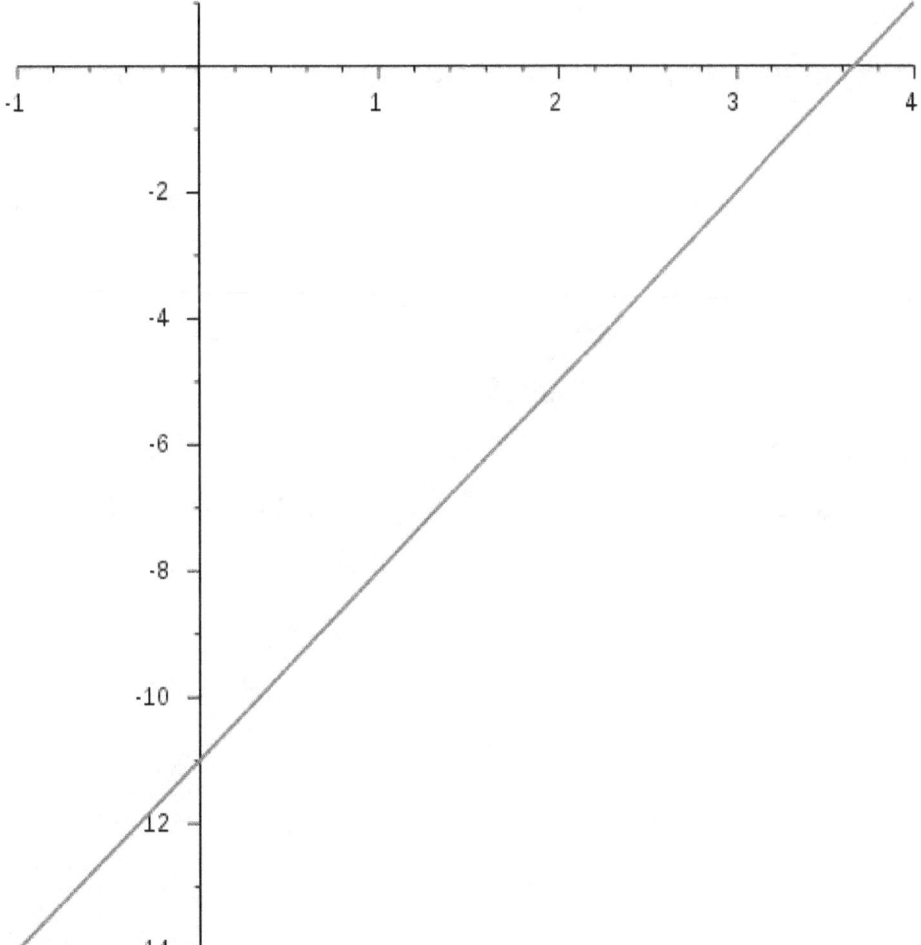

Basically, the idea here is to see at what point one side of the equation equals the other, and the easiest way to do that is to manipulate both sides until one of them is zero.

In this case, you change $3x - 4 = 7$ to $3x - 11 = 0$.

Then you set that expression as equal to y: $y = 3x - 11$. If x is 0, y is -11, and so on.

When we do this, we can simply look on the graph and see where y is 0, which in this case is where x is a bit over 3.6. If you have a graphing utility on your calculator, you can get an approximate solution much faster, but if the answer isn't a whole number, and in this case it isn't, things get a little tricky. However, while graphing isn't as much use to us here as it might be, we do have that approximate solution, which we can use to check an algebraic answer.

In general, that's what graphs are for: not a substitute for algebraic manipulation but a supplement to it. Sometimes you can get an exact solution, and sometimes you settle for a ballpark figure that will help you verify the answer you get through more painstaking work.

Now that we have the approximate answer of "a bit more than 3.6", let's do some algebra and figure out precisely what it is.

$3x - 11 = 0$

$(3x - 11) + 11 = (0) + 11$

$3x = 11$

$x = \dfrac{11}{3}$

We see that x is equal to $\dfrac{11}{3}$, or 3 ⅔, which is indeed just over 3.6. Let's try another one.

Example: $\dfrac{x}{2} + 3 = 1$

Again, shift things around a bit so we have $\dfrac{x}{2} + 2 = y$

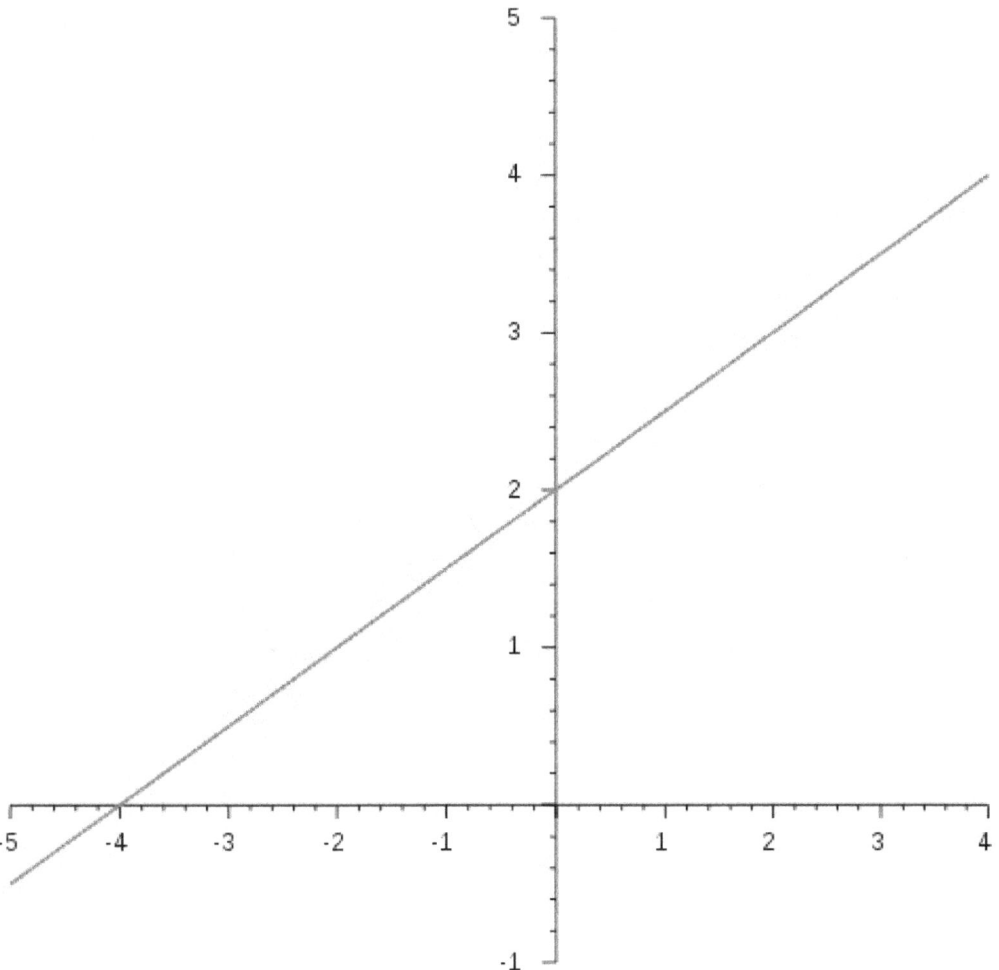

The line crosses the x-axis, and thus y is 0, at precisely where $x = -4$. An easy solution!

Let's look a little closer at this notation. We've had $y = \dfrac{x}{2} + 2$ and $y = 3x - 11$. Both are in the form $y = mx + b$, where m, the coefficient of x, is also called the **slope**, and b is the **y-intercept**.

You may have noticed, looking at those two graphs, that the lines didn't look the same: the first one, $y = 3x - 11$, was steeper than $y = \dfrac{x}{2} + 2$. That's because 3 is greater than ½ and thus the slope was steeper. Slope is also defined as the change in y divided by the change in x, when looking at two distinct points on the line, and written as $\dfrac{\Delta y}{\Delta x}$.

The Greek letter Δ, pronounced "delta", is a common marker of measuring change in mathematics. In fact, that symbol is the capital version, and the lowercase, δ, is used for a similar concept. But that doesn't come for quite a while, so don't worry about it.

One simple way to remember how to calculate slope is the phrase "rise over run": the vertical increase over (or divided by) the horizontal increase. Note that the slope can be a whole number, as in the first example, or a fraction, as in the second. It can also be a negative number, if the line sinks as you go right on the graph. Slopes are just like mountains: they can go up or down.

Example: Find the slope and y-intercept for $4x - 3 = 2x + 2$, and graph it. Then give the point where the line crosses the x-axis, that is, the coordinates on the line of $(x, 0)$.

Individual steps need not be labeled, this has been done enough by now.

$$4x - 3 = 2x + 2$$
$$(4x - 3) - 2x = (2x + 2) - 2x$$
$$(2x - 3) - 2 = (2) - 2$$
$$y = 2x - 5$$

Slope is 2, y-intercept is -5.

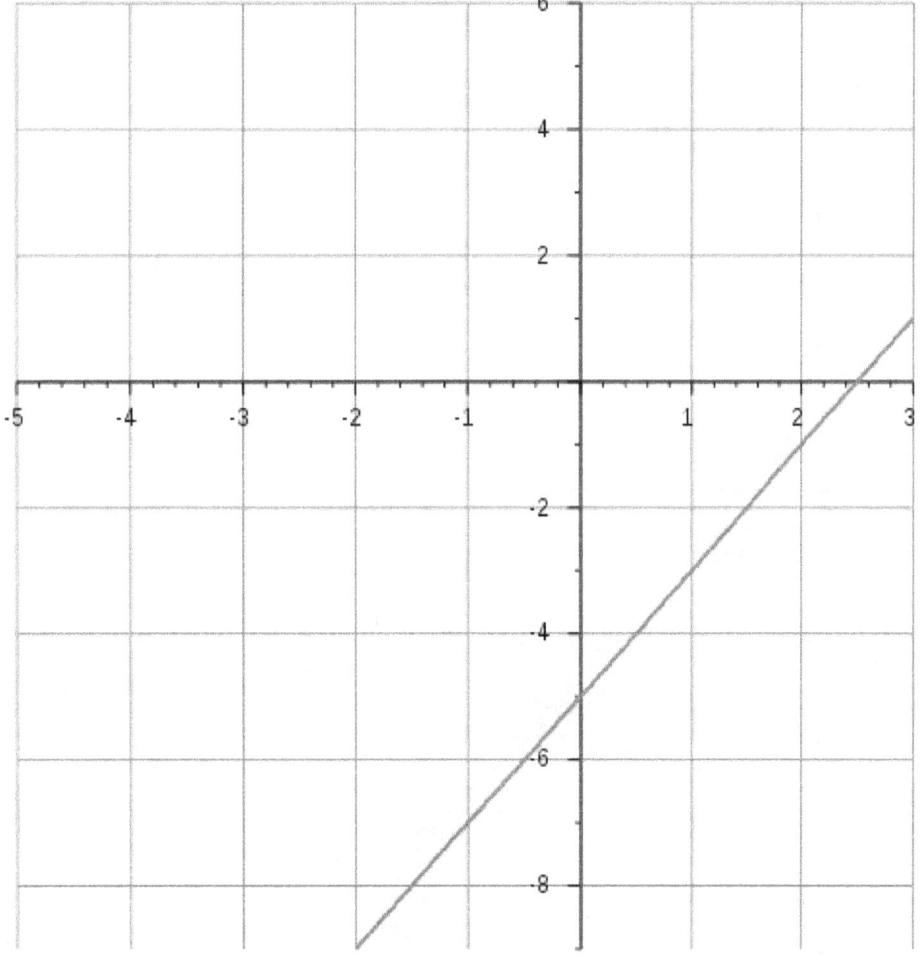

x-intercept is (2.5, 0).

Example: Find the slope and y-intercept for $\dfrac{2x}{3} + 6 = \dfrac{4x}{3} + 2$, and graph it. Then give the point where the line crosses the x-axis, that is, the coordinates on the line of $(x, 0)$.

$$\frac{2x}{3} + 6 = \frac{4x}{3} + 2$$

$$\left(\frac{2x}{3} + 6\right) - \frac{4x}{3} = \left(\frac{4x}{3} + 2\right) - \frac{4x}{3}$$

$$\left(\frac{-2x}{3} + 6\right) - 2 = (2) - 2$$

$$y = -\frac{2x}{3} + 4$$

Slope is $\dfrac{-2}{3}$, y-intercept is 4.

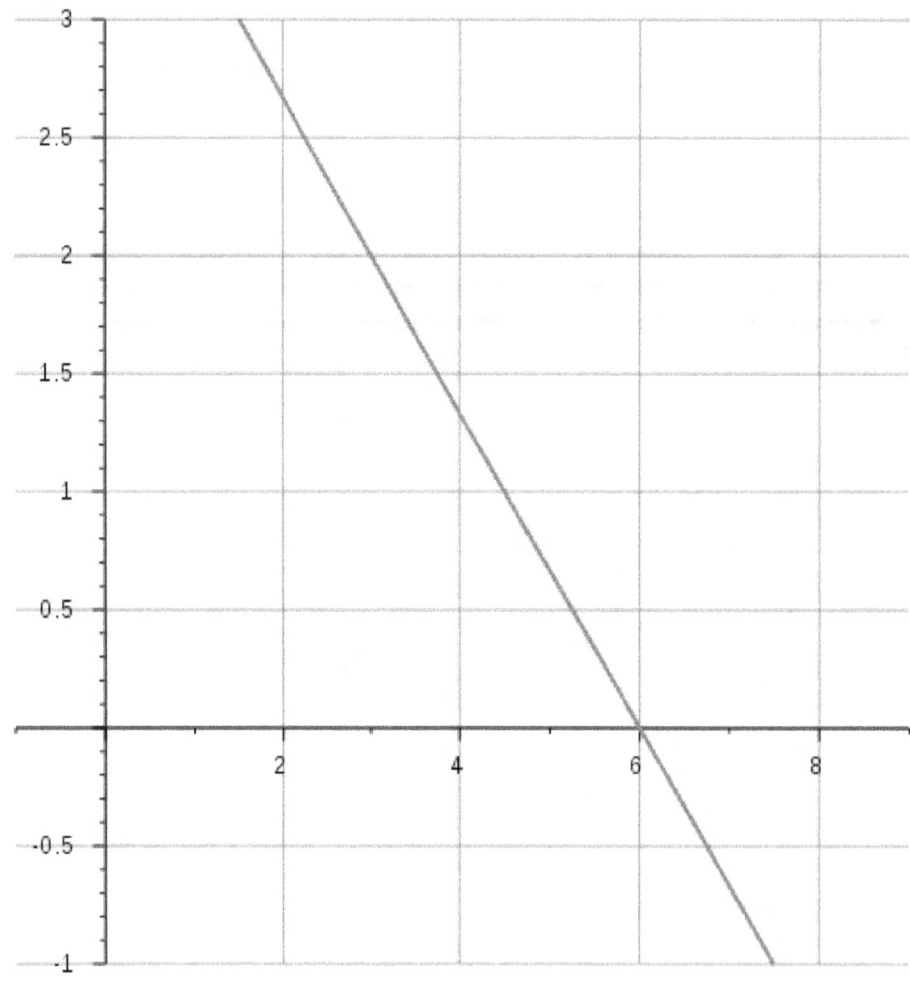

x-intercept is (6,0).

You can also solve something like $7x - 2 = 3x - 4$ by simply graphing both sides of the equation, $y = 7x - 2$ and $y = 3x - 4$, and seeing where they intersect. Mathematically, the two approaches are identical, but usually one or the other is more convenient. For example, if you're plotting revenues against expenses, you'd probably rather see how both change and where they meet (if anywhere) than simply look at the difference as a single line.

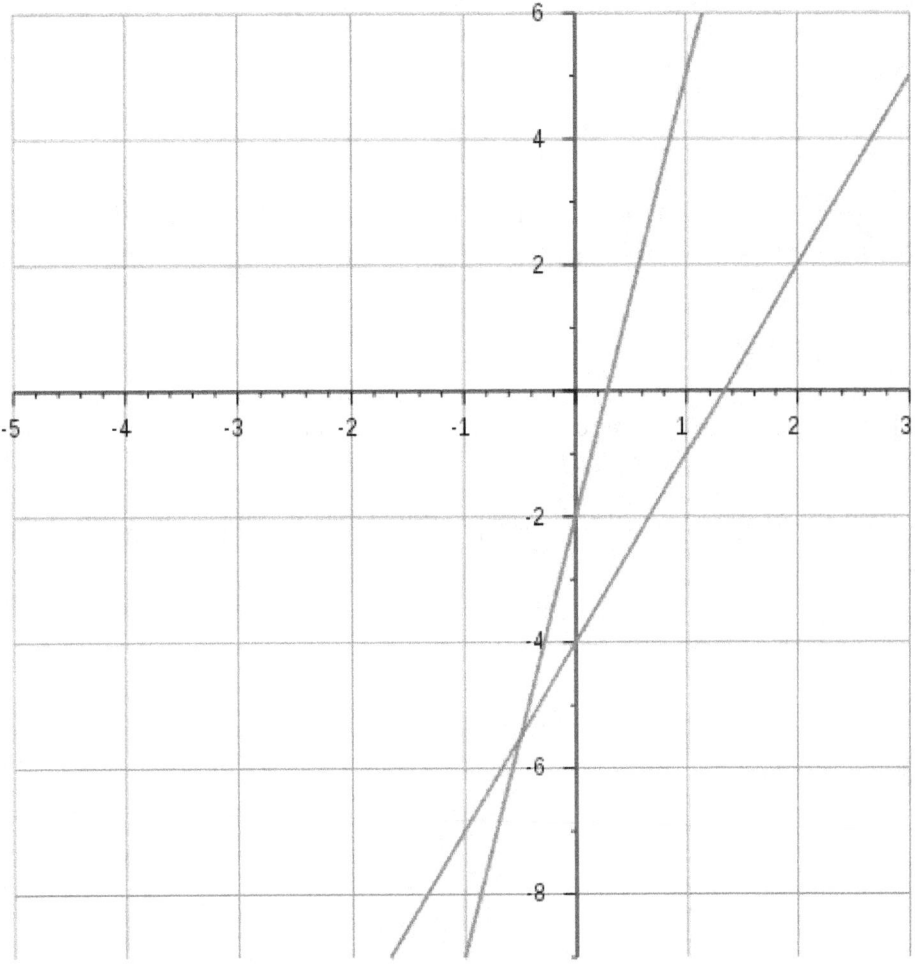

Remember that whole slope thing? Finding the slope between two points becomes important later on, so let's do it here a bit.

Find the slope between (-5,3) and (2,4).

$\Delta y = 4 - 3$

$\Delta x = 2 - (-5)$

$\frac{\Delta y}{\Delta x} = \frac{1}{7}$

Find the slope between (0,2) and (7, 5).

$$\Delta y = 5 - 2$$

$$\Delta x = 7 - 0$$

$$\frac{\Delta y}{\Delta x} = \frac{3}{7}$$

This leads us to the point-slope form of equations. It's not used as often as the slope-intercept, but it's handy if you know the slope and a point—say, if you know how steep something can be or has to be and you know where it begins or ends.

Point-slope: $y - y_1 = m(x - x_1)$

We won't work through any examples here, because all those operations have pretty much already been covered.

Lastly, graphing inequalities. Solve it as you usually do, and then plot it on either a number line or a coordinate plane.

Example: $2x - 4 < 4$

$$(2x - 4) + 4 < (4) + 4$$

$$\frac{2x}{2} < \frac{8}{2}$$

$$x < 4$$

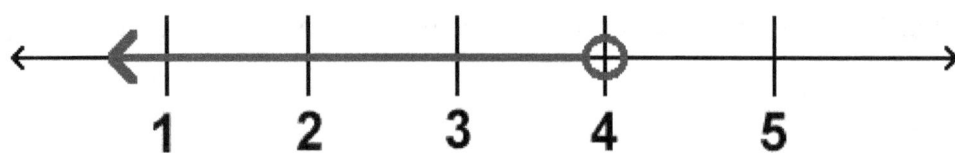

Note that the circle on the right is not filled in, because *x* cannot actually equal 4; it must be less. If $x \leq 4$, then that circle would be solid.

Example: $y \geq 3x - 2$

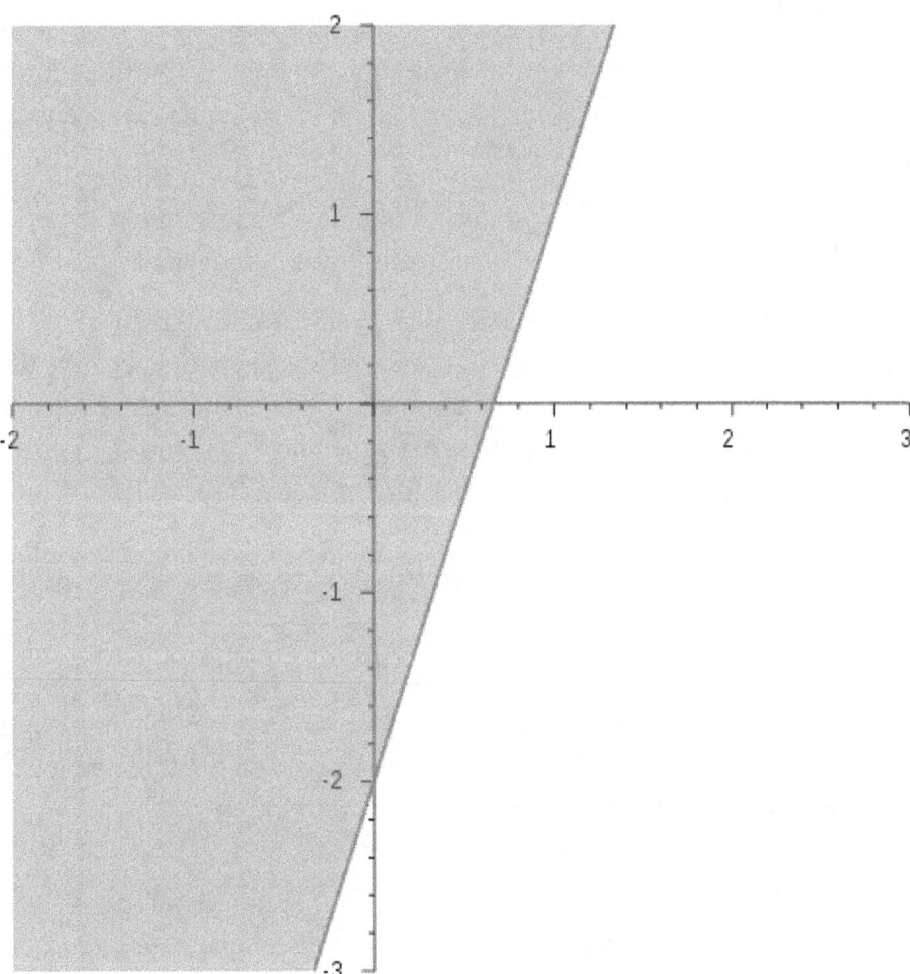

Note the darker green line on the right is solid, because y can equal $3x - 2$; otherwise, it'd be dotted.

Chapter 3: Functions

Up until now we've been talking in terms of *x* and *y*, but there's another very important notation: *f(x)*. This is pronounced "f of x", for future reference. It's largely the same thing, but there are times when *f(x)* is a lot more convenient than *y*, as we'll see later.

Every function also has two properties: the domain and the range.

The **domain** is all values for *x* that produce a valid result. For example, if your function takes the number of boxes of cookies in the cart and gives you the price, the domain is all positive numbers, because you can't have negative boxes of cookies.

The **range** is every result that the domain can produce. From the example above, the price will never be negative, even if you have coupons, so the range is also all positive numbers.

Sometimes the domain is restricted further. For example, if you're tracking the flow of water into a bucket as time passes, then while it is possible to follow the results back into negative time, it doesn't really make sense, because the amount of water isn't going to change before the faucet is turned on. Furthermore, if you measure it once per minute, then that's the range of inputs—1 cup at 1 minute, 2 cups at 2 minutes, etc., and there's no measured output for 1 minute 30 seconds.

For the material we've been looking at so far, the domain and range have both been all real numbers, positive and negative.

Example: Provide the domain and range for $f(x) = 5x + 7$.

This is just a simple linear function—that is, the only thing going on is multiplying *x* by something and adding a constant (like, say, 7). A simple linear function has a domain of all real numbers, and so does the range. That is, all real numbers will work for *x* as well as for *f(x)*.

Provide the domain and range for $f(x) = x^2 - 4$.

This is a quadratic function—that is, *x* multiplied by itself to make a square, or *quadratus* in Latin, is the base of the function, in the form $ax^2 + bx + c$. We'll see more of this in the next section.

A quadratic function also has a domain of all real numbers, but the range is not the same. Consider: if *x* is a real number, then x^2 must be positive or zero. -4 is just plain negative 4. Do the math and you will see that $x^2 - 4$ must be greater than or equal to -4. That means the range is all real numbers greater than or equal to -4, also written as $\{f(x) \in \mathbb{R} | -4 \leq f(x) < \infty\}$, or $[-4, \infty)$. Both of these mean that the output is in the set of real numbers, or rather all the real numbers that are greater than or equal to -4 but less than ∞. Less than ∞ is the standard upper bound, and it's always *less than* ∞, since no real usable number is actually infinite. If the range had been restricted to being strictly greater than -4, we would have written $(-4, \infty)$; brackets include the point in question, and parentheses do not.

One last thing about domains. Take *f(x)* = 1/x. *x* cannot equal 0, because normally in mathematics we can't divide anything by 0. So the domain is all real numbers that are not 0. The same thing applies for the range: there's no way to get *y* to equal 0, because, as *x* increases, *y* approaches but does not actually reach 0, and the same thing happens as *x* decreases. So the domain and range always start as covering all real numbers, but the details of the function may restrict them.

There's one more important property a function must have. For every input, it must produce no more than one output. That is, for functions of *x*, each *x* value in the domain has no more than one corresponding *y* value.

One easy way to check this is to graph the function. If there is a way to place a vertical line on the graph and have it intersect the function in more than one place, then we're not actually looking at a function. Intersection is important. Merely touching the function without crossing it is not enough.

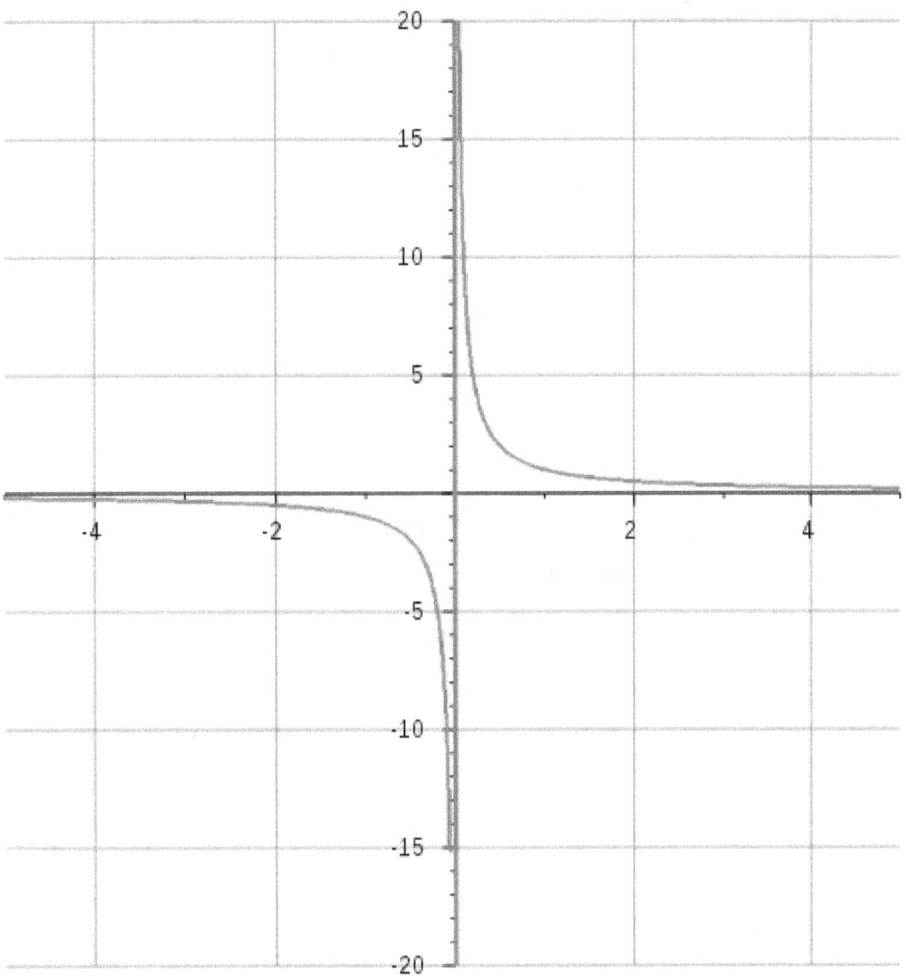

The red line ($f(x) = 1/x$) and green line ($x = 0$) touch but never cross—they don't ever actually have the same value. Thus, $f(x) = 1/x$ is a genuine function of *x* and not simply a relation, which is the function's ugly little brother that just isn't as interesting or useful.

One other topic we must cover is inversion: given *f(x)* = *y*, generate the function of *y* that gives the corresponding *x*. Now, this may seem somewhat pointless, but look at temperature conversion.

Taking a temperature in Celsius and converting it to Fahrenheit means multiplying by 1.8 and adding

$$32:$$

$$f(C) = \frac{9C}{5} + 32.$$

To get the original back, we first subtract 32 and multiply the result by $\frac{5}{9}$:

$$f^{-1}(F) = \frac{5}{9}(F - 32).$$

It's the same as solving any other equation, really, just with different symbols.

Example: Given $f(x) = 2x - 6$, produce the inverse.

Algebraic manipulations are not the point here, nor are they new, so we won't bother to write out the steps as laboriously.

$$f(x) = 2x - 6$$

$$f(x) + 6 = 2x$$

$$\frac{f(x) + 6}{2} = x$$

$$f^{-}1(x) = \frac{x + 6}{2}$$

Sometimes a function doesn't behave the same way throughout its domain—say you're buying paper and the price per box decreases for every hundred boxes you buy. It might be possible to model that as a single function, but it's much easier to say

$$f(x) \begin{cases} .25x & 0 < x < 100 \\ .20x & 100 \le x < 200 \\ .15x & 200 \le x < 300 \end{cases}$$

One other very common example is the absolute value, written as $|x|$. This function always generates a positive number with the same magnitude as the input; when x is negative, $|x|$ gives $-x$, and when it is positive the result is simply x. Again, it might be possible to model this as a single function rather than piecewise, but splitting it into those two halves is far simpler.

Chapter 4: Polynomial and Rational Functions

A polynomial function is a function consisting of a polynomial. Okay, maybe that isn't terribly helpful. All right, then a polynomial is an expression consisting of one or more terms, like $x^3 + 2x^2 - 4x + 1$. Each term must also consist of a coefficient (which can be zero) with x to the power of a positive integer, including zero. So no square roots or anything like that, just straightforward stuff like the example above.

Every term here fits in a polynomial:

$$x^5 - x^3 + \frac{2}{9}x^2 + 33x - 12$$

Each one is in the form ax^n, with n being an integer.

None of these terms do:

$$e^x - \sin x + \frac{2}{x} + x^{\frac{3}{5}}$$

The first has x as an exponent, which is not allowed. The second is a trigonometric operation, similarly not allowed. The third has a negative exponent on x, and the fourth has a fractional exponent, and neither of those is permitted either.

The most basic type of polynomial that we'll be working with is the quadratic equation, which we touched on last chapter. This is the plot of $f(x) = x^2 + 2x - 3$:

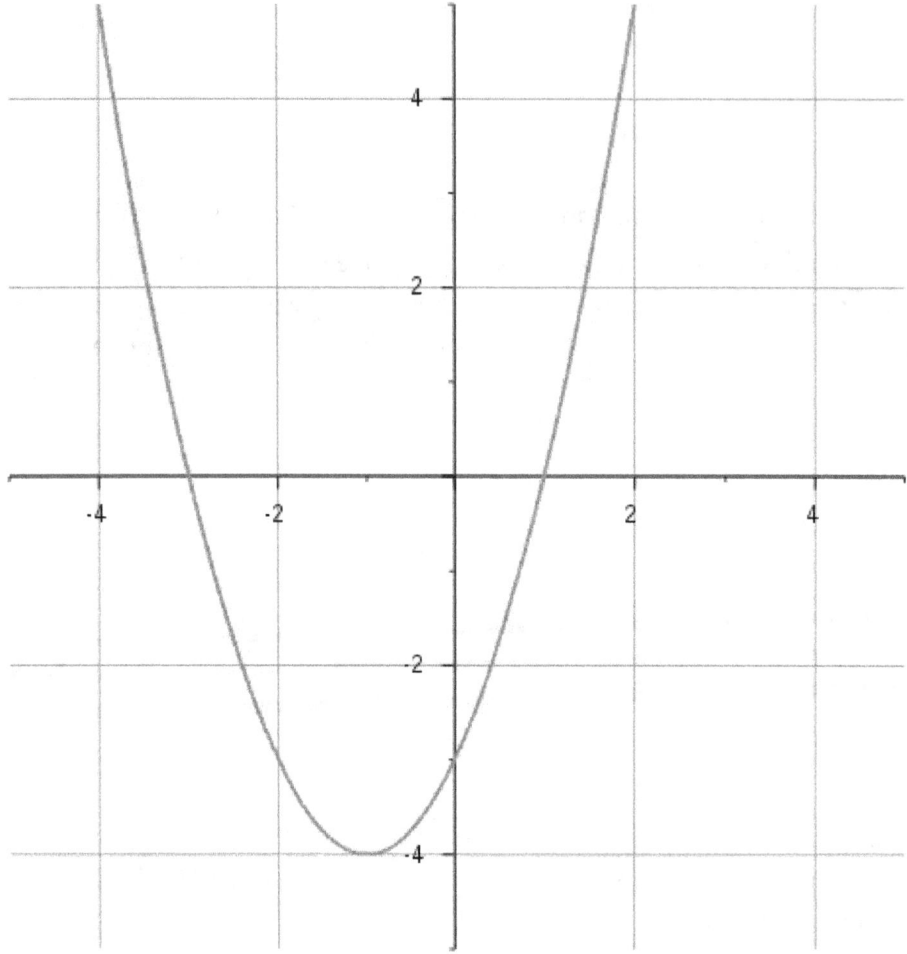

So you see here the basic shape, and a few other things. First, due to the curve, it's possible (and, in fact common) for quadratics to have two zeroes, unlike the linear equations we've worked with before. Second, with that same curve it's also quite possible for a quadratic equation to have no point that touches the *x*-axis. For example, if you moved the graph above more than 4 units up, then every point on it would have a positive *y*-coordinate, and there would be no solutions.

Monomials, binomials, and trinomials

Just so we're clear, a monomial has one term, binomials two, and trinomials three. It doesn't matter what the terms are, just how many there are.

Multiplying binomials

It's a useful thing to know how to do, so let's get to it. The primary method is FOIL—First, Outer, Inner, Last, in order of the terms you pair when multiplying two binomials. The basic idea applies for trinomials and up, but most of our work multiplying polynomials will be with binomials.

Example: Multiply $(x - 3)(x - 5)$.

The First is the terms that begin each factor—both x, so the product is simply x^2.

The Outer is the terms that bookend the whole multiplication: x and -5 here, so the product is -5x.

The Inner is the opposite, the terms on the inside: x and -3, for a product of -3x.

The Last is the terms ending each factor: -3 and -5, with a product of 15.

Summing those, we get $x^2 - 3x - 5x + 15$, or $x^2 - 8x + 15$.

Notice that the end constant, 15, is the product of the two constants in the factors, -3 and -5. Also, the coefficient of x, -8, is the sum of those same constants. We can use that for factoring quadratics as well.

Factoring quadratics

This is an important topic—if you can manage it, this is the easiest and quickest way to find the solutions of a quadratic expression. However, you do need to know how to do it, and there are a few tricks in the process, as we'll show.

Example: Factor $x^2 + 7x + 12$.

Okay, we can see immediately that this is a standard $(x + a)(x + b)$ situation, with a and b both positive and no coefficient on the x in either factor. Thus, this will be a simple factoring.

We know that the 12 is ab and the 7 is $a + b$. So all we have to do is list the factors of 12 and see which ones add up to 7.

12:

> 1 + 12 = 13
>
> 2 + 6 = 8
>
> 3 + 4 = 7

Well, that was easy. $(x + 3)(x + 4)$.

Let's try something harder.

Example: Factor $x^2 - 4x - 12$.

Hmm. Two negative signs. That means that we're dealing with one positive factor and one negative factor, and the negative factor is larger. Well, let's factor 12 again.

-12:

  1 + -12 = -11

  2 + -6 = -4

  3 + -4 = -1

2 and -6 fit the bill, giving us the required -4 coefficient for *x*. So we have

$(x + 2)(x - 6)$.

Now for the hardest kind.

Example: Factor $2x^2 + 13x + 15$.

Oh boy. $2x^2$ means that one term is $2x + a$ and the other is $x + b$; moreover, this means that while 15 is still *ab*, 13 is $a + 2b$, which is more annoying to crunch. But let's get started.

15:

  1, 15: 1 + 2 * 15 = 31, 2 * 1 + 15 = 31

  3, 5: 3 + 2 * 5 = 13, 2 * 3 + 5 = 11

Okay then, $(2x + 3)(x + 5)$.

Completing the square

All along here we've been figuring out the solutions of the equation merely by peering at the terms without really manipulating them. If you're willing to mix stuff up a bit, however, there's another method, called "completing the square". The basic idea is that instead of rendering an equation into two different factors on one side and 0 on the other, you add stuff to both sides so that they're both squares. Let's see exactly what this entails.

Example: Solve $x^2 - 4x - 3 = 0$ by completing the square.

The objective here is to turn the left side into $(x + a)^2$, or $x^2 + 2ax + a^2$. We have the $2a$ there with -4, so $a = -2$, but that's the only neat part, as -3 is definitely not a^2, which would be 4. So to make the left side a proper square, we add enough to make the last term 4, which means adding 7 to both sides.

$(x^2 - 4x - 3) + 7 = 0 + 7$

$x^2 - 4x + 4 = 7$

$(x - 2)^2 = 7$

Now, if we know that $x - 2$ multiplied by itself is 7, that means $x - 2$ can be either the positive or negative square root of 7. Remembering to include negative square roots is very important, because not all teachers give partial credit.

$x - 2 = \pm\sqrt{7}$

$$x = 2 \pm \sqrt{7}$$

The Quadratic Formula

There is a formula that will give you the solution or solutions to any quadratic equation that has a solution, which was originally worked out by completing the square for the general algebraic case of the quadratic equation.

$$\frac{-b \pm \sqrt{b^2 - 4ac}}{2a}$$

Bit of a mess, isn't it? You don't really need to know exactly where it comes from, but if you're curious the explanation is provided after the next paragraph.

Okay, we've covered polynomials pretty well, now for rational functions. And what are rational functions, you ask? Well, you know how a rational number can be expressed as the ratio of two integers? Well, a rational function can be expressed as the ratio of two polynomials. And yes, those graphs can get awfully funky. Practically everything you'd be doing algebra-wise with rational functions is the same stuff we already did, except not very much completing the square, so we won't go over that.

Appendix

Deriving the quadratic formula

We start with the general form

$$ax^2 + bx + c = 0$$

Then we start the process of completing the square by eliminating the coefficient from the first term, for

$$x^2 + \frac{b}{a}x + \frac{c}{a} = 0,$$

followed by isolating all the terms with x in them. We did this above, but in a less roundabout way.

$$x^2 + \frac{b}{a}x = -\frac{c}{a}.$$

Now we can really complete the square, by adding the constant required to have all three terms of a full square of x and a constant.

$$x^2 + \frac{b}{a}x + \left(\frac{1}{2}\frac{b}{a}\right)^2 = -\frac{c}{a} + \left(\frac{1}{2}\frac{b}{a}\right)^2,$$

Then we simply collapse the square on the left,

$$\left(x + \frac{b}{2a}\right)^2 = -\frac{c}{a} + \frac{b^2}{4a^2}.$$

and unify the fractions on the right.

$$\left(x + \frac{b}{2a}\right)^2 = \frac{b^2 - 4ac}{4a^2}.$$

We take the square root of both sides, remembering to allow for both signs

$$x + \frac{b}{2a} = \pm\frac{\sqrt{b^2 - 4ac}}{2a}.$$

and get x alone.

$$x = -\frac{b}{2a} \pm \frac{\sqrt{b^2 - 4ac}}{2a} = \frac{-b \pm \sqrt{b^2 - 4ac}}{2a}.$$

Chapter 5: Exponential and Logarithmic Functions

Now we can get into the really fun stuff: powers. Specifically, constants taken to powers of x, which is a lot quirkier (and more important for mathematics) than you might think.

The first thing we're going to talk about is one of the biggest constants in math, e. It's not a big number. In fact, it's less than 3; it's approximately 2.71828. The thing is, e is intimately involved in interest, the very first exponential function people started looking at. When interest is compounded, the bank divides the rate of interest they're giving you by the number of times per year they're compounding it and then multiplies the money in your account by that number (plus one, since 2% interest doesn't mean you retain one-fiftieth of your money) the aforesaid number of times per year. So if you have $1,000 in an account at 8% interest compounded quarterly, that thousand dollars is multiplied by 1.02 every three months, or $1000(1 + \frac{.08}{4})^4$ in a year.

Now, that's barely more than the 8% you'd get normally, but if you look at the case with 100% interest (which no bank would ever give anyone), it gets interesting. See, as the compounding interval becomes smaller, the rate of interest is divided by a larger number, but also taken to a higher power, and the resulting number does not keep increasing indefinitely. In other words, as x increases, $(1 + \frac{1}{x})^x$ approaches a certain constant, e. It shows up in other places, too, but the primary interest is in the simple function e^x. It's really tremendously useful.

Of course, it's impossible to talk about exponential functions without discussing their counterpart, logarithms. e^x raises e to the power of x, and $ln(x)$ is the power to which e must be raised to equal x. If $e^y = x$, then $ln(x) = y$, and $e^{ln(x)} = ln(e^x) = x$. They're inverse operations, and any number can be taken to the power of x, though e has special properties that make it the best base for exponentiation and logarithms. Thus, $2^3 = 8$, and $log_2(8) = 3$. 2 must be raised to the power of 3 to equal 8. Note that the $log_x(y)$ function takes the logarithm of y in base x—that is, the exponent of x you'd need to make the expression equal y. Also, we've used the $log_x(y)$ notation and $ln(x)$ as well - ln stands for natural log, which is the logarithm using base e.

Here are the graphs of e^x and $ln(x)$.

e^x:

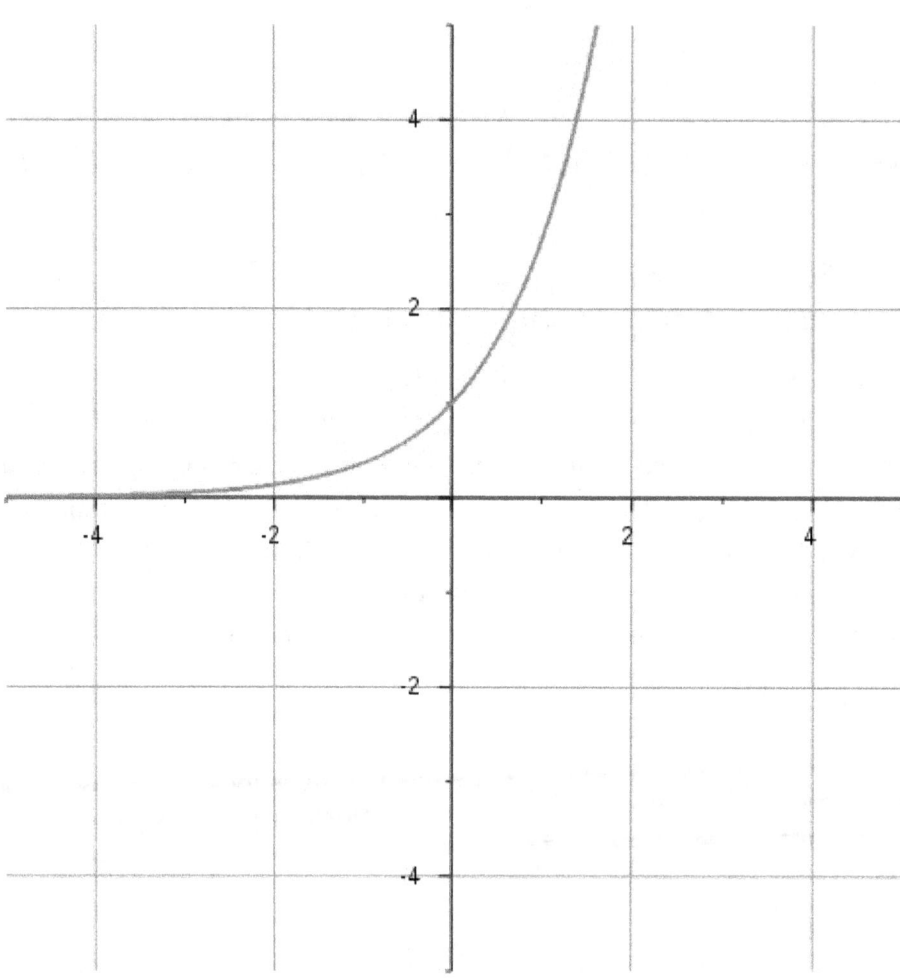

$ln(x)$:

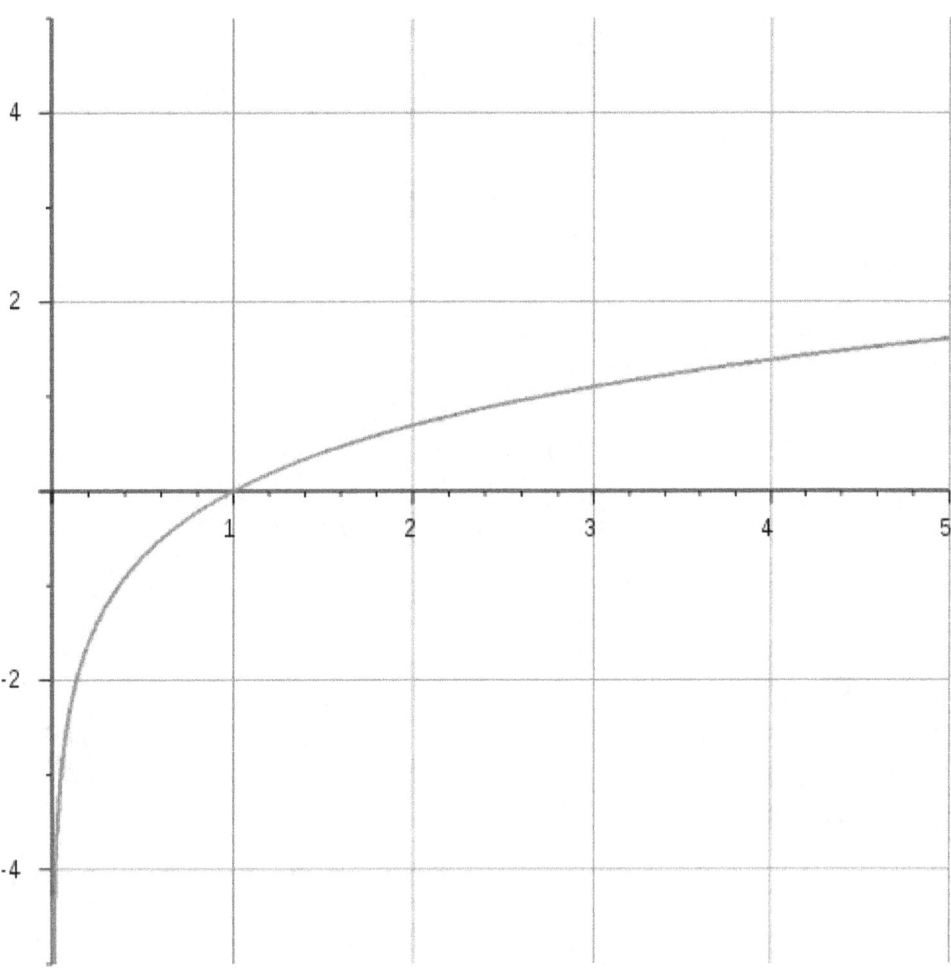

Note that e^x cannot be negative. By the same token, $ln(x)$ does not exist if $x < 0$; there's no way to make a positive number negative by taking it to any real power.

When working with exponents and exponential functions, there are a few rules to keep in mind. Using them will make things much simpler.

Rule 1: $a^0 = 1$.   Example: $2^0 = 1$.

Rule 2: $a^{-1} = \dfrac{1}{a}$.   Example: $2^{-1} = \dfrac{1}{2}$.

Rule 3: $a^{-m} = \dfrac{1}{a^m}$.   Example: $2^{-5} = \dfrac{1}{2^5}$

Rule 4: $\dfrac{a^m}{a^n} = a^{m-n}$.   Example: $\dfrac{2^4}{2^2} = 2^{4-2}$.

Rule 5: $(ab)^m = a^m b^m$.   Example: $(2 \cdot 3)^4 = 2^4 \cdot 3^4$.

Rule 6: $(a^m)^n = a^{mn}$.   Example: $(2^3)^4 = 2^{3 \cdot 4}$.

Rule 7: $a^m \cdot a^n = a^{m+n}$.   Example: $2^3 \cdot 2^4 = 2^{3+4}$

Many of these have logarithmical corollaries:

Rule 1: $a \cdot log(b) = log(a^b)$.   Example: $3 \cdot log(5) = log(3^5)$.

Rule 2: $\dfrac{log(a)}{b} = log(a^{\frac{1}{b}})$.   Example: $\dfrac{log(4)}{2} = log(4^{\frac{1}{2}})$.

Rule 3: $log(a) - log(b) = log(\dfrac{a}{b})$.   Example: $log(2) - log(5) = log(\dfrac{2}{5})$.

Rule 4: $log(a) + log(b) = log(a \cdot b)$.   Example: $log(3) + log(4) = log(3 \cdot 4)$.

All these properties are true as long as a, m, n, and b are real numbers, though the logarithmic properties require that a and b be positive. Also, the base of the logarithm must be a positive number. That's not to say that everything goes out the window and fails if stuff is negative, but the changes are beyond the scope of this guide.

There's one last operation you can perform, with both exponents and logarithms: changing the base.

With exponents: $a^x = b^{x log_b(a)}$.   Example: $2^4 = e^{4 ln(2)}$.

With logarithms: $log_a(x) = \dfrac{log_b(x)}{log_b(a)}$.   Example: $log_5(2) = \dfrac{log_3(2)}{log_3(5)}$.

In theory, any number can be used as the base of an exponential or logarithmic function or expression. In practice, there are only three that actually see widespread use: 2, e, and 10. Mathematics uses e almost exclusively, but computer science and related fields rely on 2, and biology, astronomy, and engineering are primarily 10-based. In pure algebra problems, math textbooks will retain base e most of the time, but sometimes when looking at situations in other areas, you will need to convert—so it's a good thing we displayed how to do that above.

Now, let's put some of these in action.

Example: Simplify $\dfrac{log_{10}(27) - log_{10}(3)}{2 \cdot ln(10)}$

Well, this is a bit of a mess. The first step is to unify the top, which we can do because 27 is a power of 3—specifically, the third power. There are two ways to do this, and we'll show both.

Method 1:

$$\frac{log_{10}(\frac{27}{3})}{2 \cdot ln(10)}$$

$$\frac{log_{10}(9)}{2 \cdot ln(10)}$$

Method 2:

$$\frac{log_{10}(3 \cdot 9) - log_{10}(3)}{2 \cdot ln(10)}$$

$$\frac{log_{10}(3) + log_{10}(9) - log_{10}(3)}{2 \cdot ln(10)}$$

$$\frac{log_{10}(9)}{2 \cdot ln(10)}$$

Next, we see a factor we can cancel out:

$$\frac{log_{10}(3^2)}{2 \cdot ln(10)}$$

$$\frac{2 \cdot log_{10}(3)}{2 \cdot ln(10)}$$

$$\frac{log_{10}(3)}{ln(10)}$$

The top here can be aligned with the bottom, using $log_a(x) = \dfrac{log_b(x)}{log_b(a)}$.

$$\frac{\frac{ln(3)}{ln(10)}}{ln(10)}$$

$$\frac{ln3}{ln(10) \cdot ln(10)}$$

There are a number of ways we can further fiddle with this, but those tend to eliminate one factor at the expense of messiness, so we'll leave it here.

Example: $\dfrac{log(4) + log(25)}{ln(e^2)}$

Well, here's some simplifying we can do right off the bat: $e^{ln(x)} = ln(e^x) = x$

$$\frac{log(4) + log(25)}{2}$$

And we can combine those logs, too:

$$\frac{log(4 \cdot 25)}{2}$$

Um, yeah. We'll combine them all the way:

$$\frac{log(100)}{2}$$

And using the same rules as above:

$$\frac{2}{2}$$

Ooh boy, that'll be a doozy.

1

Back before calculators were widespread, people used tables of logarithms for work and had to look them up individually—ship navigators, cannon operators, and others. There were books on logarithms, hundreds of pages of numbers, and the entire thing was a pain in the neck, but it was all people had. In fact, it's only fairly recently that math textbooks stopped doing that, and they still tend to include logarithms up to 100 or so.

Before 1650 or so, those guys had to do all that arithmetic by hand, instead of saving effort by adding, subtracting, dividing, or multiplying logarithms. If that sounds like a enormous pain, that's because it was, even more than not being able to use a computer. Quick, who wants to multiply out $729 \cdot 2458$ rather than adding $2.863 + 3.39$ and looking up the answer? Yeah, that's what I thought.

Now that we've covered exponents and logarithms, let's go to something completely different in the next chapter.

Chapter 6: Systems of Equations and Inequalities

As we said before, now we come to something completely different: systems of linear equations and inequalities. The first thing we're going to cover is a real world application: groceries. A car trunk after a trip to the store contains 20 cans weighing 100 ounces. Some of the cans are cat food, 3 ounces each, and the rest are chunky soup, 11 ounces each. How many of each kind of can is in the trunk?

Okay, we've got two equations here, with a being the number of cat food cans and b as the number of chunky soup cans. We know how many ounces they each weigh, how many total cans there are, and the weight of the cans in the trunk.

Equation 1: $a + b = 20$

Equation 2: $3a + 11b = 100$

There are several ways to solve this system, two of which we'll be looking at in this part of the section. The first is solving one equation for a single variable—in this case, turning equation 1 into $a = 20 - b$—so we can plug that into the other equation: $3(20 - b) + 11b = 100$. Thus we can give b an actual value, 5, with a bit of elementary algebra, and that in turn tells us that $a = 15$. So there are 15 cat food cans and 5 chunky soup, or $(15, 5)$ as our **solution set**.

Let's try another one. Two medium-sized books and one textbook weigh 8 pounds. That textbook weighs 2 pounds more than one medium-sized book. How much does each book weigh?

Equation 1: $2x + y = 8$

Equation 2: $y - x = 2$

Here we'll be using the other method: isolating a variable by adding and subtracting equations. Examining this pair, we can either obtain the value of y by adding Equation 2 to Equation 1 twice, or do the same for x by subtracting Equation 2 from Equation 1. Either way, we can use the answer to determine the value of the other variable, so it doesn't really matter which one we pick.

Therefore, we have:

$(2x + y) - (y - x) = (8) - (2)$

$2x + y + x - y = 6$

$3x = 6$

$x = 2$

Plugging that into Equation 2, for simplicity's sake, we get $y - 2 = 2$, or $y = 4$, making the solution set $(2, 4)$.

Okay, we've covered the context and why systems of linear equations are useful. Now let's go over the requirements for solving them.

First, every equation must be independent—if you have two equations and one is a multiple of the other, then it's really just a single line. Similarly, if you have three equations and one can be obtained by manipulating the other two, then you can take it out and not lose any information. Note that these can still be solved—if two lines are the same, then the solution set is infinite, and in the latter case you might have three lines that intersect at one point, or three planes that intersect on a line.

Second, they must be consistent, meaning that they have a solution. If two equations are parallel lines, then obviously there is no point where they meet. We can often find this by a little equation manipulation: if the sum or difference or whatever leads to a logical contradiction like equating two different constants, then you can stop there. Also, if there are three equations with just x and y, then each pair of lines might meet in a different place, with no common solution.

Got it? Good. Let's do some more complex ones.

Example:

Equation 1: $2x - y = -1$

Equation 2: $y - 4x = -3$

Adding 1 and 2 will cancel out y, for $-2x = -4$, $x = 2$, and using that for Equation 1, $4 - y = -1$, $y = 5$, so the solution set is $(2, 5)$.

Still pretty simple. Let's add a third variable: z, which is customary for denoting the axis of depth, or height when looking top-down, as in that case y is usually depth and x is breadth.

Equation 1: $4x + 3y - 3z = 7$

Equation 2: $x - y + 2z = 4$

Equation 3: $2x + y + z = 11$

Now this is more like it. A common strategy to simplify the situation is juggling stuff around until you have two equations dealing with the same two variables, which you can easily use together to solve for both.

Eq. 1-2*Eq. 3: $y - 5z = -15$

Eq. 3-2*Eq. 2: $3y - 3z = 3$

That second one can be reduced to $y - z = 1$ and rearranged to use in the first as $y = z + 1$, for $z + 1 - 5z = -15$. That in turn reduces to $z = 4$, and thus $y = 5$. Plugging those into any of the three original equations yields $x = 1$, for an intersection point of $(1, 4, 5)$.

There's one last situation you need to be aware of, where the solution set is a bit different, but you can't just say, "System has no solution, done!" If there are more variables than equations, then one variable per equation must be defined in terms of the rest; if you have x, y, and z, but only two equations, then two of them are solved as depending on the third.

Example:

Equation 1: $x + y - 2z = 1$

Equation 2: $-x + 2y + 5z = 8$

It doesn't really make any difference which variable you choose to treat as independent, since you'll be describing the same line in the end, just using different terms. Just as before, then, the general goal will be to make it as easy as possible—take whatever stands out, avoid fractions if possible because they're annoying.

In this case, we can see right off the bat how to get rid of x (and thus state y in terms of z or vice versa): simply add the equations. That gives us $3y + 3z = 9$, or $y = 3 - z$. Substituting that, in turn, into the first equation (since that involves less multiplication) yields $x + (3 - z) - 2z = 1$, or $x = 3z - 2$. Since we have only two equations, that's as far as we can go, a line in three dimensions with z free and the other two defined from it.

Lastly, we can deal with systems of inequalities, though they don't yield to algebraic methods—it's an enormous pain to define an area where all three (or more) sides are lines with defined non-zero slopes (that is, no vertical or horizontal lines). Thus, the usual solution is graphing, which in turn means that inequalities are virtually always given in two dimensions, since graphing in three is quite difficult.

Example:

Equation 1: $y - 2x + 1 < 3$

Equation 2: $y + 2x - 3 \leq 0$

Equation 3: $y - x > -1$

Shifting those around a bit, we get $y < 2x + 2$, $y \leq -2x + 3$, and $y > x - 1$, which produces this graph, with the two sides on the left being dotted and the one on the right solid.

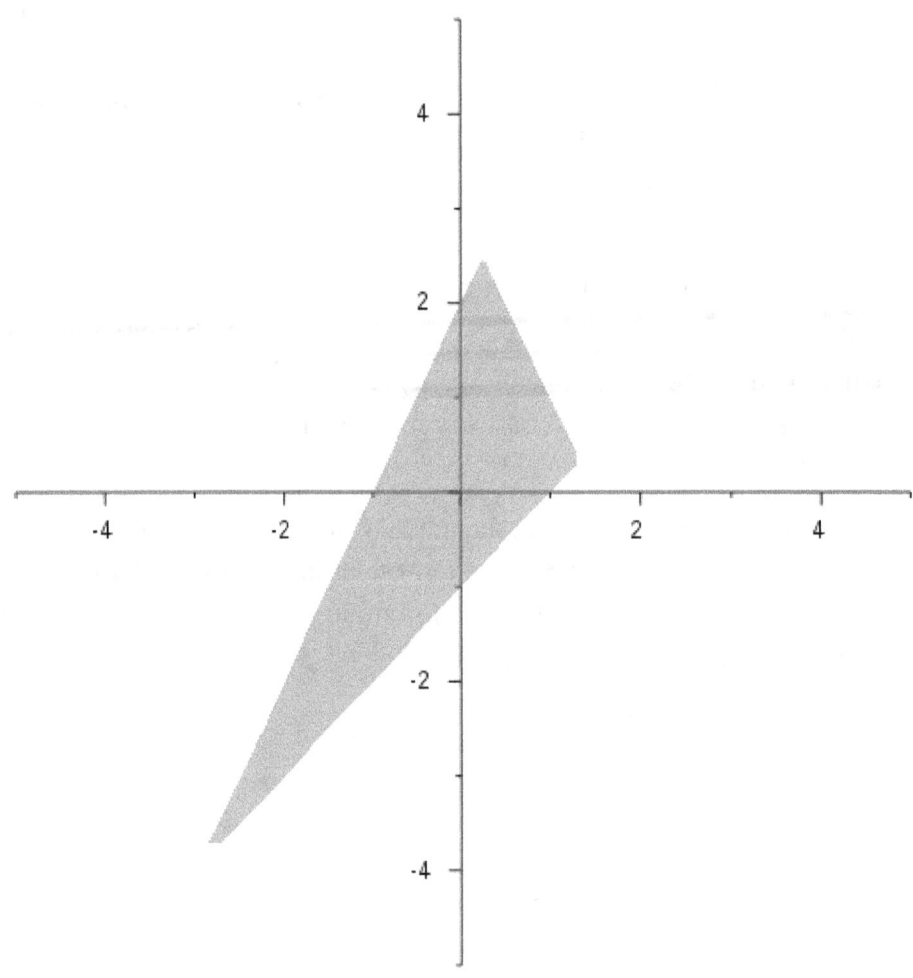

38

Chapter 7: Matrices and Determinants

Now we come to the last method of solving systems of equations, though we won't be using it much for that in this chapter: matrices. A matrix, in case you haven't encountered them before, is a computer simulation keeping humans asleep while the machines harvest our energy—um, a rectangular array of numbers, variables, or expressions; we'll mostly be dealing with the first two.

Here's what we're talking about, basically:

$$\begin{bmatrix} 5 & 8 \\ 1 & 3 \end{bmatrix}$$

There are a number of things we can do with matrices. Addition and subtraction are normal, multiplication and division are significantly different, and there are three new operations: inversion, transposition, and taking the determinant. Let's cover multiplication first.

Matrices A and B can be multiplied $A \cdot B$ only if A has the same number of columns as B has rows. This means that matrix multiplication is not commutative: $A \cdot B$ is not necessarily equal to $B \cdot A$, if either is possible. Let's see how matrices are multiplied. (At the end of this chapter, you will have seen the word "matrices" that it is not going to look like a word.)

$$\begin{bmatrix} 1 & 4 \\ 5 & 8 \end{bmatrix} \cdot \begin{bmatrix} 3 & 7 \\ -2 & 6 \end{bmatrix} = \begin{bmatrix} 1 \cdot 3 + 4 \cdot -2 & 1 \cdot 7 + 4 \cdot 6 \\ 5 \cdot 3 + 8 \cdot -2 & 5 \cdot 7 + 8 \cdot 6 \end{bmatrix} = \begin{bmatrix} -5 & 31 \\ -1 & 83 \end{bmatrix}$$

We see here how it works. To determine the entry in row i, column j, multiply each member of the first matrix's row i with the corresponding member of the second matrix's column j. That, by the way, is why the first matrix must have the same number columns as the second has rows. That's the only way for the first matrix to have the same length row as the second matrix's column. Without that, there's an unpaired entry and things get screwy.

Okay, that's multiplication. From there, we go to finding the determinant, then inversion, then division, and you'll see why we chose that order as we finish each section.

Finding the determinant

The determinant of the matrix $\begin{bmatrix} a & b \\ c & d \end{bmatrix}$ is $ad - bc$. 3x3 matrices are, unfortunately, considerably more complicated, but they will be on pretty much any test involving this material, so here's how to do it. There are actually two ways to write the formula, and we'll include them both—you get the same answer no matter what, and it stands to reason that different ways of looking at it will be easier for different people.

$$M = \begin{pmatrix} a_{11} & a_{12} & a_{13} \\ a_{21} & a_{22} & a_{23} \\ a_{31} & a_{32} & a_{33} \end{pmatrix},$$

The matrix in question is labeled , since that shows the row and column structure better.

Method 1: multiplying diagonals.

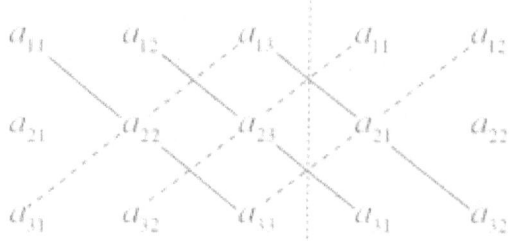

Multiply and add each trio of terms linked by a solid line, and subtract from that sum the products of each dashed line. That is,

$$a_{11}a_{22}a_{33} + a_{12}a_{23}a_{31} + a_{13}a_{21}a_{32} - a_{13}a_{22}a_{13} - a_{11}a_{23}a_{32} - a_{12}a_{21}a_{33}.$$

Method 2: mini-determinants.

Take each element of the top row in turn. Cross out the terms in its row and column, and multiply it by the determinant of the resulting 2x2 matrix. The first and last terms of the top row are positive, the second is negative.

$$a_{11}\begin{vmatrix} a_{22} & a_{23} \\ a_{32} & a_{33} \end{vmatrix} - a_{12}\begin{vmatrix} a_{21} & a_{23} \\ a_{31} & a_{33} \end{vmatrix} + a_{13}\begin{vmatrix} a_{21} & a_{22} \\ a_{31} & a_{32} \end{vmatrix} =$$

$$a_{11}a_{22}a_{33} + a_{12}a_{23}a_{31} + a_{13}a_{21}a_{32} - a_{13}a_{22}a_{13} - a_{11}a_{23}a_{32} - a_{12}a_{21}a_{33}$$

Got that? Good. Now, to invert a matrix, calculate the determinant, and divide each term by it. If the determinant is 0, the matrix is not invertible.

Example: Invert $\begin{bmatrix} 3 & 5 \\ 2 & 7 \end{bmatrix}$.

The determinant is simply $3 \cdot 7 - 5 \cdot 2 = 11$, and then divide it through for $\begin{bmatrix} \frac{3}{11} & \frac{5}{11} \\ \frac{2}{11} & \frac{7}{11} \end{bmatrix}$. A matrix times its inverse is the identity matrix, which in 2x2 form is $\begin{bmatrix} 1 & 0 \\ 0 & 1 \end{bmatrix}$, and more generally has a value of 1 when the column number and row number are equal, and 0 otherwise. This matrix generally has the properties for matrices that 1 has for ordinary numbers—multiplying or dividing yields the original number, and (as we just said) dividing a matrix by itself gives the identity matrix.

Okay, let's try something a little trickier.

Example: The determinant of $\begin{bmatrix} 5 & 0 & -1 \\ 3 & 1 & 0 \\ 0 & 5 & x \end{bmatrix}$ is -5. Find x.

Okay, maybe more than a little trickier, but really, this shouldn't be too hard.

The first step is to start calculating the determinant. We prefer the second method described above, but again, either works.

$$5 \cdot \begin{vmatrix} 1 & 0 \\ 5 & x \end{vmatrix} - 0 \cdot \begin{vmatrix} 3 & 0 \\ 0 & x \end{vmatrix} + -1 \cdot \begin{vmatrix} 3 & 1 \\ 0 & 5 \end{vmatrix} = -5$$

$$5 \cdot (1 \cdot x - 0 \cdot 5) - 0 \cdot (3 \cdot x - 0 \cdot 0) + -1 \cdot (3 \cdot 5 - 1 \cdot 0) = -5$$

$$5 \cdot (x) - 1 \cdot (15) = -5$$

$$5x = 10$$

$$x = 2$$

See? Not too bad. Now for division (Note that a matrix must have the same dimensions of row and column to be invertible; otherwise, you just can't do it, like multiplying a 2x2 matrix and a 3x3).

Division is really just more of the same. To divide Matrix A by Matrix B, invert B and multiply A by it. That's AB^{-1}, not $B^{-1}A$—they can be the same, but there's no guarantee of it. In fact, they usually aren't.

Remember last chapter, about solving systems of equations? Well, you can use matrices for that, though it's rather cumbersome, and you need to know how because it does get tested. So here's how it works.

$$a_{11}x + a_{12}y + a_{13}z = b_1$$

$$a_{21}x + a_{22}y + a_{23}z = b_2$$

$$a_{31}x + a_{32}y + a_{33}z = b_3$$

There's the system in algebraic form. Now here's the solution.

$$x = \frac{\begin{vmatrix} b_1 & a_{12} & a_{13} \\ b_2 & a_{22} & a_{23} \\ b_3 & a_{32} & a_{33} \end{vmatrix}}{\begin{vmatrix} a_{11} & a_{12} & a_{13} \\ a_{21} & a_{22} & a_{23} \\ a_{31} & a_{32} & a_{33} \end{vmatrix}} \quad y = \frac{\begin{vmatrix} a_{11} & b_1 & a_{13} \\ a_{21} & b_2 & a_{23} \\ a_{31} & b_3 & a_{33} \end{vmatrix}}{\begin{vmatrix} a_{11} & a_{12} & a_{13} \\ a_{21} & a_{22} & a_{23} \\ a_{31} & a_{32} & a_{33} \end{vmatrix}} \quad z = \frac{\begin{vmatrix} a_{11} & a_{12} & b_1 \\ a_{21} & a_{22} & b_2 \\ a_{31} & a_{32} & b_3 \end{vmatrix}}{\begin{vmatrix} a_{11} & a_{12} & a_{13} \\ a_{21} & a_{22} & a_{23} \\ a_{31} & a_{32} & a_{33} \end{vmatrix}}$$

Basically, the numerator is the coefficients of all the variables in matrix form, with the coefficients of the variable in question replaced with the constant expression in each equation. The denominator is the same thing but without that replacement. Thus, it is the same for all variables. This method extends to more complex systems, but we won't be covering that here. 3x3 matrices are enough of a pain in the neck.

Let's try this out with a 2x2 system, since it's the same principle and vastly less annoying. Note that this method works only when there are as many variable as equations, since otherwise the matrices won't be square and determinants won't exist.

Equation 1: $2x - y = 1$

Equation 2: $x + y = 5$

$$x = \frac{\begin{vmatrix} 1 & -1 \\ 5 & 1 \end{vmatrix}}{\begin{vmatrix} 2 & -1 \\ 1 & 1 \end{vmatrix}}$$

The first set of determinants, for x:

$$y = \frac{\begin{vmatrix} 2 & 1 \\ 1 & 5 \end{vmatrix}}{\begin{vmatrix} 2 & -1 \\ 1 & 1 \end{vmatrix}}$$

The same for y:

So $x = \dfrac{1 - (-5)}{2 - (-1)}$ and $y = \dfrac{2 \cdot 5 - (1)}{2 - (-1)}$, which simplifies to $x = \dfrac{6}{3}$ and $x = \dfrac{9}{3}$, or $(2, 3)$, which checks out with the equations above.

You should always check your answers, by the way. We haven't covered this in the past, but it's important, along with checking your work. It's really irritating to lose some or all of the points available because your theory was sound but you made a mistake with arithmetic or because you just transcribed things incorrectly. Plugging the values that you've calculated into complex equations can be awfully annoying, but again, it's a lot better than being moved down a letter grade because you missed something. And it doesn't matter how good you are at math—everyone makes mistakes. It takes everybody a while to get used to going over problems once they were solved, but it's worth it if you catch even a couple of errors.

Chapter 8: Conic Sections

Up until now we've been dealing mostly with algebraic treatments of the topics we covered. Only graphs and inequalities really needed proper pictures. This chapter, however, will require a lot of visualization, and so we'll begin with images before we start equations.

We all know what a cone is, right? The shape that sand, flour, sugar, salt, etc. form if you just pour them out? A rounded version of the pyramids? Well, mathematically the full form, which is two cones stacked vertically, points together, has some very interesting properties, as do four shapes derived from cutting into it.

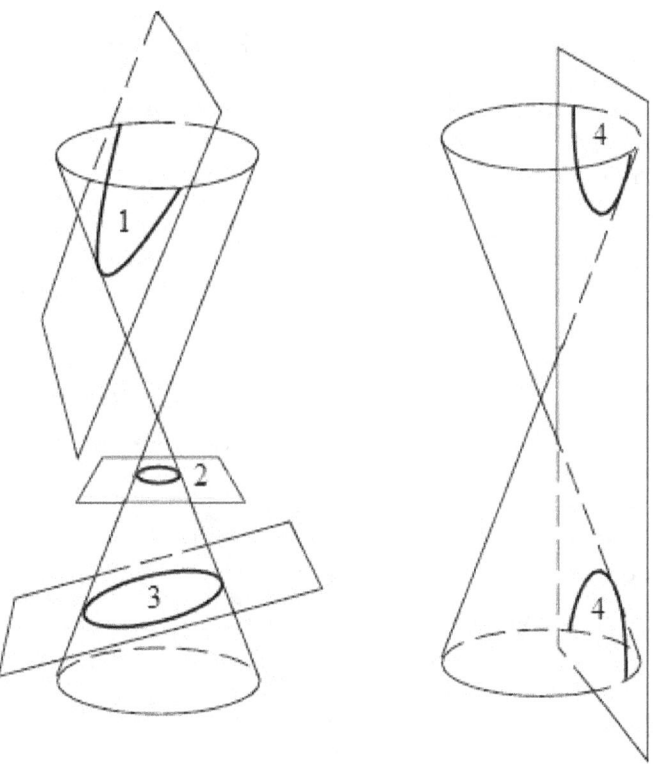

Example 1 is a **parabola**, which we've seen before. We'll be looking at the equation in a slightly different way this time, though.

Example 2 is a **circle**, because the plane that cuts the cone is flat. Example 3 is an **ellipse**, because the plane in question is tilted with relation to the cone. An ellipse is a circle that has been stretched in one direction.

The last one is the trippiest, the **hyperbola**. It consists of two disconnected shapes described by a single equation. We've already run into this a bit with $f(x) = 1/x$, and we'll go into more depth on it later.

Parabolas

There are three components to every classic parabola: the line, the point, and the curve. The line is called the **directrix**, the point is the **focus**, and the curve is every point that is the same distance from each.

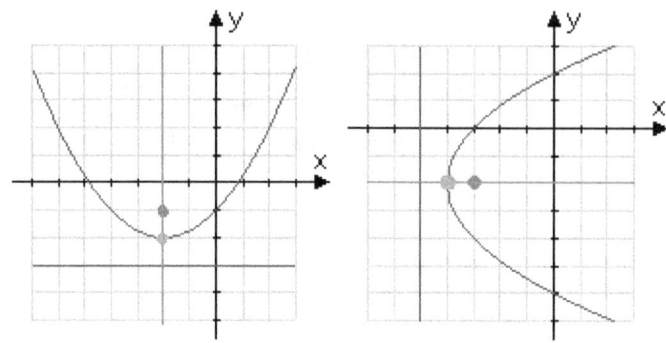

Note that parabolas can be functions of x or y—in fact, they can be functions of both at once, though the algebra gets really hairy there.

The equation describing the first graph you know as $y = ax^2 + bx + c$, with the orange dot, also known as the **vertex**, at the point $\left(\frac{-b}{2a}, \frac{-b^2}{4a} + c\right)$. There's also another form of that equation: $y = a(x - h)^2 + k$ and conversely $x = a(y - k)^2 + h$ with the vertex at (h, k). And a third form: $4p(y - k) = (x - h)^2$. $4p$ is just another term for $\frac{1}{a}$; it's replaced because p is the distance between the vertex and the directrix or focus - both intervals are the same, by definition. Be prepared to recognize and deal with both forms, because you may encounter either.

Let's look at an example: $y = 4(x - 2)^2 + 3$. The vertex is at $(2, 3)$, and since $a = 4$, $4p = \frac{1}{4}$, and $p = \frac{1}{16}$. Since this is an ordinary quadratic with the independent variable positive, the arc opens up, so the focus is $\frac{1}{16}$ above the vertex, or $\left(2, 3\frac{1}{16}\right)$, and the directrix is the same interval below, at $y = 2\frac{15}{16}$. The axis of symmetry, which we haven't mentioned previously, is always perpendicular to the directrix and runs through the vertex, and in this case that makes it simply $x = 2$.

Chapter 9: Circles and ellipses

The next most complicated shapes, and the only ones that are self-contained, are the circle and ellipse. A circle, as you probably know, is the set of all points that are distance r from the center (h, k).

General equation, which needs to be converted to something else to be of any use:

$$x^2 + y^2 + Dx + Ey + F = 0$$

Center-radius form, which is much handier:

$$(x - h)^2 + (y - k)^2 = r^2$$

Ellipses are much the same, except stretched in one direction. Ellipses also have two foci to a circle's single center. Also, there are vertices on the outer tips of the short sides (that is, far apart), and co-vertices on the other outer tips (that is, closer together). The distance from the center, between the two foci, to the far vertex is a, while the distance from the center to the near co-vertex is b, and from the center to either focus is c:

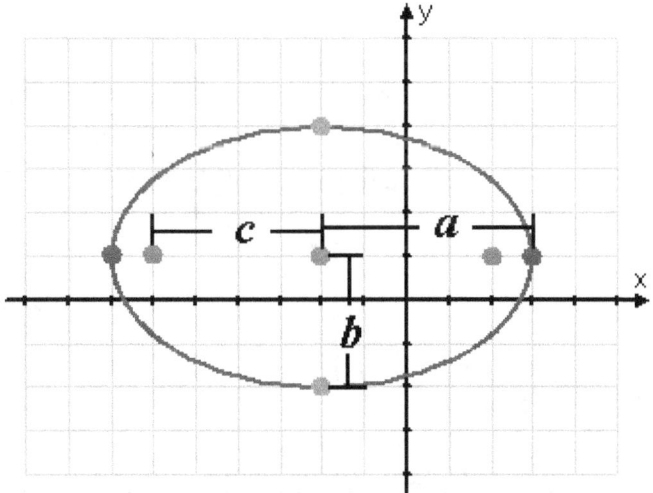

$b^2 + c^2 = a^2$, which is awfully silly because all of this is based on the Pythagorean theorem, and you'd think they would have kept the letters from that to make it easier to remember, but so it goes. We'll just have to be careful.

Now, for the relevant equations so you can actually work with these.

For a wide ellipse with its center at (h, k), vertices a units to the right and left of the center, and each focus c units right and left from the center:

$$\frac{(x-h)^2}{a^2} + \frac{(y-k)^2}{b^2} = 1$$

For a tall ellipse, with its center at (h, k), vertices a units above and below the center, and each focus c units above and below the center:

$$\frac{(y-k)^2}{a^2} + \frac{(x-h)^2}{b^2} = 1$$

Ellipses also have the concept of eccentricity—not being rich and weird, but a measure of how much the ellipse is stretched away from being perfectly circular. This is defined as $e = \frac{c}{a}$; the closer the foci are to the center, the smaller the eccentricity, though there must be some or it is just a circle. Note that $c < a$, since the focus is always closer to the center than the vertices are, and thus $0 < e < 1$.

Ellipses are more complex circles, so we'll look at one of them: $\frac{(x-3)^2}{5^2} + \frac{(y+2)^2}{3^2} = 1$. This is a wide ellipse, since the largest denominator is under x; the center is at $(3, -2)$—remember that it's $(y-k)^2$, so we're really looking at $(y-(-2))^2$. Since $a = 5$ and $b = 3$, $c = 4$. Thus, the left focus is at $(-1, -2)$ and the right is at $(7, -2)$; eccentricity is .8.

Hyperbolas

Here we come to the tricky part. Up until now, we've had shapes formed from points equidistant from a point and a line, a point, or two points. Now we have a figure composed of the points where the *difference* in the distance from two points is equal, instead of the sum of that distance.

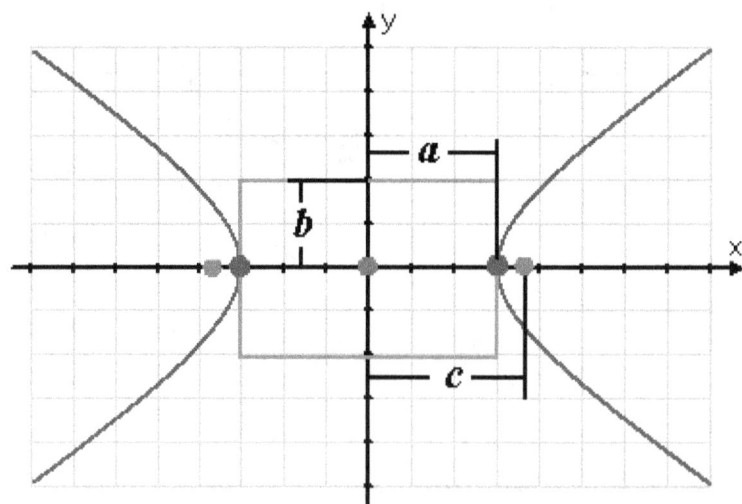

Unlike an ellipse, the foci of a hyperbola are further from the center than the vertices are.

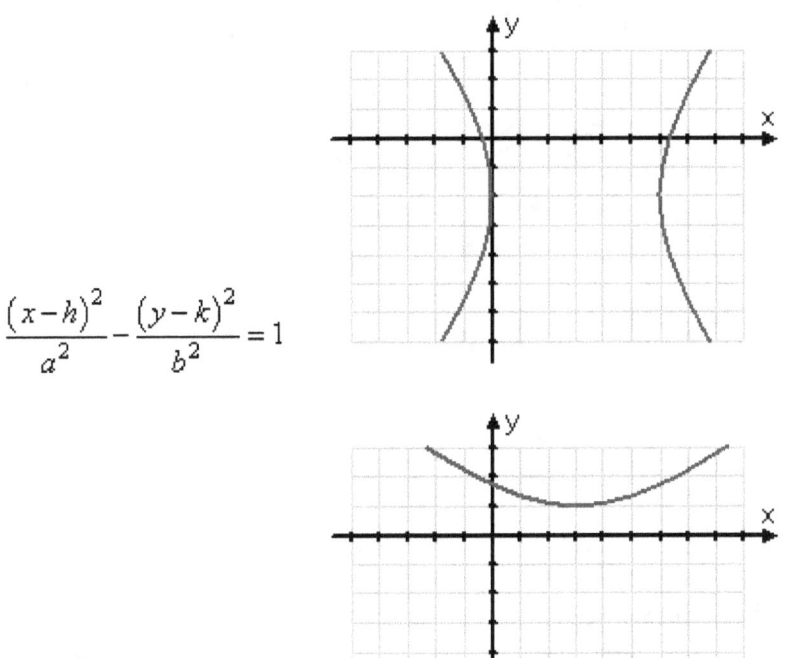

$$\frac{(x-h)^2}{a^2} - \frac{(y-k)^2}{b^2} = 1$$

$$\frac{(y-k)^2}{a^2} - \frac{(x-h)^2}{b^2} = 1$$

If you zoom out on these a bit, you'll see that both branches approach a pair of lines but never meet them. The **asymptotes** are an important part of parabolas, and you'll often have to identify them.

Here are the formulas for each set, like above:

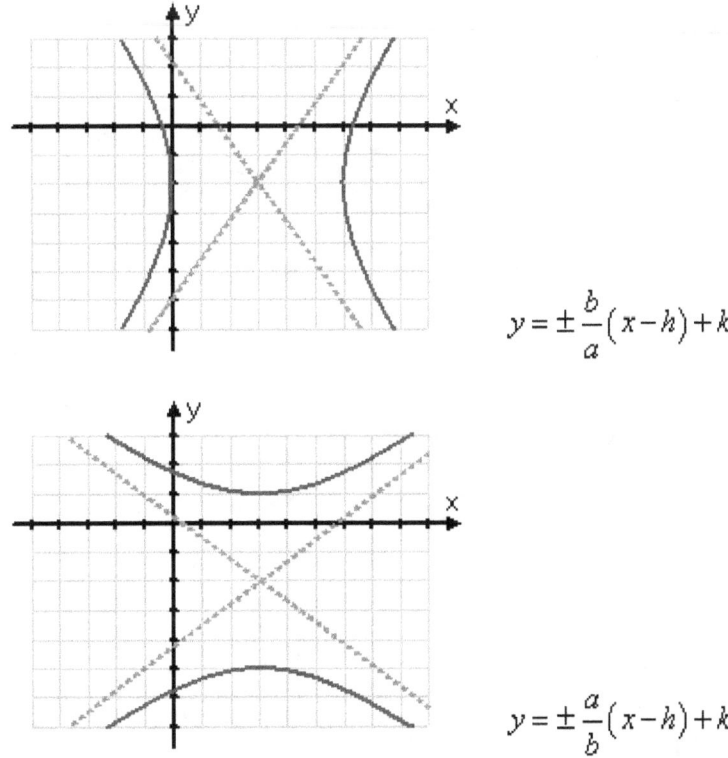

$$y = \pm \frac{b}{a}(x-h) + k$$

$$y = \pm \frac{a}{b}(x-h) + k$$

Eccentricity is relevant here as well. It is defined as $e = \dfrac{c}{a}$ again, with $b^2 + c^2 = a^2$. Since the foci are further from the center than the vertices are, $c > a$, and thus $e > 1$. A higher eccentricity means that the hyperbolas are closer to being flat, like a gentle speed bump, and lower eccentricity approaches the appearance of a spike.

Example: $\dfrac{(x-1)^2}{5^2} - \dfrac{(y+1)^2}{3^2} = 1$

x is positive, so the branches open left and right, and the center is $(1, -1)$. The square root of the denominator of the positive term is 5, so $a = 5, b = 3$, and $c = 4$ again. The right focus is $(5, -1)$, while the left is $(-3, -1)$, and the asymptotes are $y = \pm\dfrac{3}{5}(x-1) - 1$, or $y = \dfrac{3x}{5} - 1\dfrac{3}{5}$ and $y = \dfrac{-3x}{5} - \dfrac{2}{5}$.

The last two chapters will be quite different from the rest and even from each other. However, this one is less independent than the next, so let's get started.

Chapter 10: Sequences and Series

A sequence is a list of numbers—or objects, or anything with definable characteristics, but here we'll only be looking at numbers—generated by a formula or otherwise predictable. Every sequence that isn't just random numbers has a rule to it. Much of this section will be about deducing that rule and using it to calculate more terms of that sequence. Here's an example.

Example: Sequence A consists of the terms 1, 4, 9, 16 . . . Give the next three terms.

All of these terms are squares, so in all likelihood the rest of them are as well, and the next three terms are the next three squares, 25, 36, 49.

We said "in all likelihood" because given any number of points, there will always be infinitely many functions that include them all. The sequence 2, 3 could be simply $x + 1$ for $x = 1, 2$, or it could be $-\dfrac{x^2}{2} + \dfrac{5x}{2}$ for $x = 1, 2$, or $\dfrac{x^3}{7} + \dfrac{13}{7}$ on the same range, or any number of others. There's really no way to know for sure, but generally we go for the simplest answer that is correct—in the case above, $S_n = n + 1$. That, by the way, is one of the notations you'll need to learn—S_n is term n in sequence S.

Example: Sequence B consists of the terms 1, 3, 7, 13 . . . Give the formula for the series and the next term.

This, now, is a bit trickier. Nothing comes out at you immediately, so let's go to the default method: finding the common difference. If the difference between terms is constant, then the sequence is linear; if the difference between terms is not constant but increases or decreases at a constant rate (in other words, taking the differences between terms as their own sequence, and the difference between those terms is constant), then the sequence is quadratic, and so forth. If that's not going anywhere, try calculating the ratio between terms—if that's constant, then the series is exponential. Naturally, there are many other functions you can use to generate a sequence, but if you're given several terms and told to deduce the formula behind them, you'll pretty much always be dealing with $an^2 + bn + c$ or ax^n.

Anyway, the actual example. Calculating the difference between terms, we have 2, 4, 6, which is not constant but is increasing at a constant rate, so this is fundamentally quadratic. To solve it the rest of the way, we'll use a system of equations. We know term 1 is 1, and $an^2 + bn + c$ with $n = 1$, and similarly for the rest of the terms, so we have three unknowns and four terms to deduce them from.

Equation 1: $1a + 1b + c = 1$

Equation 2: $4a + 2b + c = 3$

Equation 3: $9a + 3b + c = 7$

Standard system, $3a + b = 2, 2a - c = 1, 11a + 3b = 8$. We end up combining the first and third for $a = 1$, using that in the first for $b = -1$, and in the second for $c = 1$. So $S_n = x^2 - x + 1$.

Okay, so that's sequences. Now, series are pretty similar—instead of a list of terms, you add all those terms up. Obviously, if those terms are greater than 1 and increasing, that'll be an infinite sum, so most of the series we'll be discussing are decreasing fractions that approach a finite sum.

Example: Provide the eventual sum of $1 + \dfrac{1}{2} + \dfrac{1}{4} + \dfrac{1}{8} + \cdots$.

Well, this has even less of an immediate solution. How do we handle it?

We can tell that each term is half the last, with the first simply being 1, so $S_n = \dfrac{1}{2^{(n-1)}}$, beginning with $n = 1$. But the sum of those terms is something else entirely, and it's time to introduce a new symbol.

$\sum$ means "sum", and more specifically $\displaystyle\sum_{k=0}^{n} ar^k$ means the sum of all terms ar^k, from 0 to whatever n is (usually infinite). Believe it or not, there is a simple way to state that sum, composed as it is of infinitely many terms.

First, let's write it out a bit:
$$\sum_{k=0}^{n} ar^k = ar^0 + ar^1 + ar^2 + ar^3 + \cdots + ar^n.$$

Next, multiply it by $(1 - r)$, for reasons that will become clear eventually:

$$(1 - r)\sum_{k=0}^{n} ar^k = (1 - r)(ar^0 + ar^1 + ar^2 + ar^3 + \cdots + ar^n)$$

$$= ar^0 + ar^1 + ar^2 + ar^3 + \cdots + ar^n$$
$$- ar^1 - ar^2 - ar^3 - \cdots - ar^n - ar^{n+1}$$
$$= a - ar^{n+1}$$

After that restatement, we divide both sides by $(1 - r)$, restoring the left side to its original form:

$$\sum_{k=0}^{n} ar^k = \frac{a(1 - r^{n+1})}{1 - r}.$$

Now, if we're dealing with a geometric series containing decreasing terms, then $r < 1$, and if n "increases without bound", to use the formal terminology, then r^{n+1} is effectively 0 and $1 - r^{n+1}$ is

$$\sum_{k=0}^{n} ar^k = \frac{a}{1 - r}.$$

just 1, making the whole thing simply . Remember this formula, because you're going to use it in pretty much any problem involving series.

Now, back to the example. We already determined that each term was the previous term multiplied by $\frac{1}{2}$,

so that's the value for r, and the first term is 1, so that's the value for a, giving us $\sum_{k=0}^{n} \frac{1}{2}^{k} = \frac{1}{1 - \frac{1}{2}}$, which comes out to 2.

Oh, since we haven't stated this before—any series that has a finite sum (that is, adding the terms, usually by way of the formula above, yields a number) is convergent, and any series that does not (say, the one in the first example, which increased rapidly) is divergent.

Now, series can be added and subtracted, and even multiplied, though the vast majority of problems involving series will just be addition and multiplication. As always, let's show it through an example.

Example: Give the sum of $1 + \frac{3}{7} + \frac{-1}{3} + \frac{2}{7} + \frac{1}{9} + \frac{4}{21} + \frac{-1}{27} \ldots$.

What on Earth is going on there? At first glance, these terms don't seem to go together—because they don't. This is the sum of two series, with the terms mixed together. What we're really looking at here is $(1 + \frac{-1}{3} + \frac{1}{9} + \frac{-1}{27} + \ldots) + (\frac{3}{7} + \frac{2}{7} + \frac{4}{21} + \ldots)$. The first series has $a = 1$ and $r = -\frac{1}{3}$, while

the second has $a = \frac{3}{7}$ and $r = \frac{2}{3}$. In both cases the absolute value of the ratio between terms is less than 1 (multiplying by -5 repeatedly diverges no less than multiplying by +5), so we can use the formula

above to calculate the sum—or sum of sums, if you will. The first series comes out to $\sum_{k=0}^{n} \frac{1}{1 - (-\frac{1}{3})}$,

or $\frac{3}{4}$, while the second is $\sum_{k=0}^{n} \frac{\frac{3}{7}}{1 - (\frac{2}{3})}$, or $\frac{9}{7}$, for a total of $\frac{57}{28}$.

Chapter 11: Probability

Now we come to the most disconnected part of this guide and one that you most likely will use more than all the others combined: probability. We rely on it for all predictions, especially weather—though we do joke about how inaccurate those predictions often are. In this section, you're going to learn a bit about how probability works.

First of all, the terminology. A situation being examined is called a **experiment**, and every possible ending of that situation is an **outcome** individually and the **sample space** when considered as a group. When computing the probability of an event, you divide the number of favorable outcomes by the total number of outcomes.

Example: What is the probability of rolling a 4 or better on one die?

Answer: There are six outcomes possible, and three of them are greater than or equal to 4. Thus, the probability is $\frac{3}{6}$, or 50%.

Note that this applies only if every outcome is equally likely. If some have a better chance of happening than others, then you must weight your calculations accordingly.

There are several qualities that events may have. For example, since if you roll a die, you will only get one outcome, all die results are **mutually exclusive**: that is, if one happens, none of the others can. In addition, the set of all the outcomes is **collectively exhaustive**, since it accounts for all possibilities. The **complement** of an event is the opposite of that event, the chance that it does not occur. Lastly, if the occurrence of event A does not affect the probability of event B, they are **independent**. All of these are important for performing calculations and analyzing situations.

Example: Calculate the chance of drawing a red card or a club from a deck of cards.

Solution: On a single draw, you cannot get a card satisfying both conditions of being red and a club. Thus, they are mutually exclusive, and the probability of this event is simply the sum of both events. 26 out of 52 cards are red, and another 13 are clubs, so the probability of drawing a red card or a club from a deck of cards is $\frac{39}{52}$.

That last problem wasn't quite fair, since it used a concept we haven't introduced yet: combining probabilities. There are two major words that we'll be using here, *and* and *or*. *And* means that the event must satisfy both conditions, while *or* means that one, or the other, or both can be satisfied.

There's also something called a **compound event**, which is just like it sounds: two or more events being considered as a group. This can be two different things (like rolling a die and flipping a coin) or more than one of the same thing (like rolling five dice or rolling one die five times).

Example: Calculate the probability of rolling 1 at least once with six rolls of a six-sided die.

Solution: This is a little trickier than it looks. At first, you might think it's simply a matter of six separate rolls, and since only one roll needs to come up 1, it's a simple sum of 6 rolls, like with the red card or club at the beginning. However, if you take the chance of rolling 1 once and multiply it by 6, that gives a probability of 100%—and while it's not likely that you will roll six times and never get a 1, it's certainly not impossible. We're going to use another property of events here, complementing: the chance of rolling 1 at least once with six rolls plus the chance of never rolling 1 with six rolls is equal to 100%. The only way to accomplish the latter is a combination of six rolls that all must be something other than 1, which means multiplying $\frac{5}{6}$ by itself six times, instead of adding $\frac{1}{6}$ to itself six times. That probability is 33.5%, which means the opposite, seeing at least one 1 on the die, is 66.5%.

Now, let's deal with a larger sort of situation. If you've got a red shirt, a white shirt, and a blue shirt, in how many ways can you arrange them? Well, you can choose any of the three shirts for the first slot, so there are three possibilities there. The second slot has two options, since you've already picked one shirt, so there are two possibilities. The third slot has just the one choice, the only shirt left. So mathematically, three possibilities and two possibilities, using the rules of probability above, multiply the chances for six total outcomes. That seems right, but let's check it. RWB, RBW, BWR, BRW, WBR, WRB. It does work: six arrangements in total. This is called a **permutation**, and it also introduces the concept of the **factorial**.

Put simply, the factorial of n, denoted as $n!$, is the product of n with every integer smaller than itself. (Strictly speaking, the function also works for non-integers, but we're not covering that here.) The factorial of n is also the number of ways you can put n objects in order.

But wait, there's more! What if you've got six shirts, all different colors—how many different ways can you choose three for spending a few nights at a friend's house? Well, the calculation is pretty similar: six choices for the first slot times five for the second times four for the third means 120 ways total. But that counts shirt 1, shirt 2, shirt 3 as different from shirt 2, shirt 1, shirt 3, and most of the time you're not going to care what order they're in. If that's the case, you're looking at a **combination**. That is the case here, so we take those 120 ways and divide by the number of ways to order three objects, which as we calculated above is 6, to get 20 different sets of 3 shirts, picking from 6. Combinations are also referred to as a choose b, or like here, 6 choose 3.

Combination notation:
$$C(n,r) = {}^nC_r = {}_nC_r = \binom{n}{r} = \frac{n!}{r!(n-r)!}$$

Permutation notation:
$$P(n,r) = {}^nP_r = {}_nP_r = \frac{n!}{r!}$$

Conclusion

Well, it's been an interesting tour through the topics of algebra. We hope you found this guide helpful, since, after all, that's why we wrote it. Good luck in your future math courses, and remember: math doesn't have to be boring. With mathematics, as with everything else, it's all in how you approach it. Even if you're not naturally talented at math, you can still comprehend the material if you look at it the right way and put in some effort. Ultimately, that's what it comes down to sometimes, and hard work is necessary throughout life. Good luck!

About Minute Help Press

Minute Help Press is building a library of books for people with only minutes to spare. Follow @minutehelp on Twitter to receive the latest information about free and paid publications from Minute Help Press, or visit minutehelpguides.com.

www.ingramcontent.com/pod-product-compliance
Lightning Source LLC
Chambersburg PA
CBHW081305170526
45165CB00011B/3413